PRAISE FOR *EINSTEIN ON THE ROAD*

"With warm sympathy and a rare ability to grasp both modern science and twentieth-century history, Josef Eisinger introduces us to an Einstein who was touchingly engaged in coming to grips with a world that by turns puzzled, enchanted, moved, idolized, and threatened him. *Einstein on the Road* opens a wide window on the great physicist's mind and his times."

—Jerrold E Seigel, Kenan Professor of History Emeritus
at New York University and author of *Bohemian Paris*

"This book presents a fascinating summary of Einstein's unpublished travel diaries, covering the crucial decade between his emergence as the supreme scientific superstar (1922) and his permanent emigration to the United States (1933). The narrative places many known events of Einstein's life into a new context and also presents personal tidbits unknown even to those who consider themselves Einstein experts! For those, as well as the wider readership, the book also contains a rich set of footnotes with surprising connections to the worlds of science, politics, and culture."

—Pierre Hohenberg, professor of physics, New York University

EINSTEIN ON THE ROAD

EINSTEIN ON THE ROAD

Josef Eisinger

Prometheus Books

59 John Glenn Drive
Amherst, New York 14228-2119

Published 2011 by Prometheus Books

Cover photo by Philippe Halsman © Halsman Archive
Cover design by Nicole Sommer-Lecht

Inquiries should be addressed to
Prometheus Books
59 John Glenn Drive
Amherst, New York 14228–2119
VOICE: 716–691–0133
FAX: 716–691–0137
WWW.PROMETHEUSBOOKS.COM

15 14 13 12 11   5 4 3 2 1

Library of Congress Cataloging-in-Publication Data

Eisinger, Josef, 1924–
     Einstein on the road / Josef Eisinger.
          p. cm.
     Includes bibliographical references and index.
     ISBN 978–1–61614–460–9 (cloth : alk. paper)
     ISBN 978–1–61614–461–6 (ebook)
     1. Einstein, Albert, 1879–1955—Diaries. 2. Einstein, Albert, 1879–1955—Travel.
I. Title.

QC16.E5E586 2011
530.092—dc23

                                                                      2011019045

Printed in the United States of America on acid-free paper

To Styra

Contents

Foreword

Walter Gratzer

Was there ever a public figure like Einstein? With a few strokes of the pen, any cartoonist could bring him to life on the page. His features would probably have been familiar to a Pennsylvania miner and to a rickshaw driver in Delhi. His features still adorn billboards and advertising logos. When Einstein was on the road, crowds would turn out to greet him or merely to goggle, flashbulbs would pop, and microphones would be thrust in his face. His scientific confrères viewed him with astonishment and awe; the public, with veneration; and Hitler's regime, since he was both a pacifist and a Jew, with loathing. "Relativity" entered the vernacular and became the generally uncomprehended and misused subject of both dinner-table conversation and jokes (Einstein leans out of a train window in London and asks a porter, "Does Oxford stop at this train?"). When in 1919 his theory of relativity was proclaimed correct, following Eddington's measurements of the deflection of light from the sun's disc during an eclipse, it made headlines in newspapers around the world. Lord Haldane, twice Lord Chancellor of Great Britain, brother and uncle of illustrious scientists, even informed the head of the Church of England that the theory would have important theological implications. The archbishop, alarmed, procured a heap of books on the subject, and his attempts to grasp its meaning had driven him close to panic and despair when he had the good fortune to encounter its progenitor at Haldane's house. Einstein was happily able to reassure him that "relativity is a purely scientific matter, and has nothing to do with religion," thus relieving the distraught prelate's torment.

Why then did Einstein, a man who never sought—who indeed shunned —publicity, and yearned for solitude, so impose himself on the public imagination? In part, no doubt, it was his physiognomy: the disheveled hair, the bright, soft eyes, betokening deep thought; his mind, like that of Newton (whose picture hung in his study), "forever voyaging through strange seas of thought, alone." Einstein was moreover something of an enigma. A vignette of his brief time at Oxford comes from the classicist Gilbert Murray, who found him one day sitting in the great quad of his college, Christ Church, "with a far-away look on his face." "The far-away thought behind that far-away look was evidently a happy one, for, at that moment, the exile's countenance was serene and smiling." "Dr. Einstein, do tell me what you are thinking," Murray demanded. "I am thinking," Einstein answered, "that after all, this is a very small star."

Einstein was, besides, a citizen of the world, or so he certainly regarded himself and was regarded by others: he was one of them. Nationalism he thought "an infantile sickness. It is the measles of the human race." He once mused that should his theory of relativity prove true, "Germany will claim me as a German and France will declare I am a citizen of the world. Should my theory prove untrue, France will say I am a German and Germany will declare I am a Jew." (Not, of course, that he remotely doubted its validity.) Even his rabid enemies in Germany hesitated to disown him altogether. To them, Einstein observed, he might be a "stinking flower," but they still wanted to put him in their buttonhole (see chapter 4).

Then there was the mesmeric mystery of relativity: when Arthur Eddington was asked by a reporter whether it was true that there were only three men in the world who could understand it, he asked who the third might be. And, in a counter to Alexander Pope's famous couplet, "Nature and Nature's Laws lay hid in night. / God said 'let Newton be,' and all was light," the man of letters Sir John Squire wrote, "It did not last: the Devil howling 'Ho! / Let Einstein be!' restored the status quo." This was indeed the public perception in a nutshell.

It is no surprise, then, that Einstein has been the subject of more biographies than any other scientist—perhaps even any other man who ever lived. Some of them focus on his science, some on his personal life, one even on his (abundant, it has to be said) love life. And yet there is one area of the territory that has eluded, until now, the biographers and miners of archives. Josef Eisinger, himself a distinguished physicist who understands the theory of relativity, has laid hands on the meticulous diaries that Einstein kept throughout the period of his extensive travels. In ceaseless demand, Einstein journeyed, most often with his wife, Elsa, from continent to continent and country to country, lecturing on physics, on pacifism, on the political catastrophes of the time, sometimes to raise money for the infant Hebrew University in Jerusalem, sometimes for pacifist causes. Plagued by endless demands for interviews, forced to submit to the hated tuxedo and to attend dinners, receptions, and award ceremonies, all of which he abhorred, yet often discovering interesting people and old friends, Einstein recorded his experiences and his feelings and opinions with characteristic candor. In between public appearances, he would toil in hotels and in ships' cabins at the elusive unified field theory to which he devoted the last three decades of his life. Ever and anon he would reach out to seize it by the tail, only to have it slip from his grasp yet again. His greatest consolation was his violin: wherever he found himself, he would seek out musicians, amateur and professional, for a rapturous evening of Mozart, Haydn, or Brahms quartets and sonatas, which would leave him elated and refreshed.

Most striking, perhaps, is Einstein's ceaseless curiosity, the unrelenting activity of his mind. He ponders the local vegetation, the birds and fishes, the reflections on the sea, the light penetrating the fog. When his ship encounters a storm, and Elsa and the other passengers take to their bunks, Einstein stands on the bathroom scales, observes that his weight oscillates between maxima and minima in the ratio of 3:2, and calculates the acceleration of the vessel when it drops into the trough between two waves. But there is more to this book, for Einstein became intimate with many of the most celebrated political, artistic, and intellectual figures of the time, recording his impressions and opinions of them. We are witness to a pageant of remarkable characters, both famous and obscure, who flit through its pages, all against a backdrop of a scientific revolution and of shifting political sands. Josef Eisinger illuminates his text with engrossing endnotes of unfailing interest and authority about men, women, and events in a turbulent and momentous period of history. Read and learn.

Walter Gratzer,
Professor Emeritus,
King's College, London

Foreword

Peter Lax

Great scientists are very seldom public figures or culture heroes. Their achievements are so far removed from ordinary life and intuition that the general public cannot appreciate them nor give them much thought. Albert Einstein is the modern exception; everybody knows his name and recognizes his face; in his lifetime he was an object of adulation whenever he appeared in public; the prominent and famous were eager to meet him.

There is no doubt that Einstein richly deserved his uncommon fame. He was one of the most significant and most original physicists of the twentieth century. A complete unknown in 1905, he burst upon the world of physics with three papers that have completely changed our view of space, time, and energy. However, his public fame came much later, in 1919, when an expedition to Brazil created a sensation by confirming Einstein's theory of the deflection of light by the gravitational field of the sun.

Most good scientists follow their own ideas and ignore the opinion of the scientific community; Einstein had that trait in the extreme. For instance, when all leading physicists hailed the new quantum mechanics, he refused to believe that it was the ultimate theory, because he would not accept the uncertainty principle.

In the second half of his scientific career, he was an isolated figure; the following story, told to me by the mathematician Max Schiffer, illustrates the extent of his isolation. Einstein went to the director of the Institute of Advanced Study, Robert Oppenheimer, and suggested that the institute invite some talented young physicist interested in the general theory of relativity to spend a year or two there. Oppenheimer's reply was: "Professor Einstein, no talented young physicist is interested in the general theory of relativity."

Einstein disregarded many social conventions, not only in dress and appearance, for neither he, nor Schrödinger, his close associate when both lived in Berlin, held monogamy in high esteem, as the author of this book points out.

The volume you hold in your hand is a wonderful, novel kind of biography. You accompany Einstein on his trips, hear what he has to say, meet the people he met. You learn of his love of music—how even on his busiest days he found time to play chamber music; professionals were eager to play with him. The intimacy of this account of Einstein's daily life is matchless. The notes for each chapter offer interesting and often surprising information about Einstein's historical times—for example, General Ludendorff's view of Hitler in 1933 (see note 3 to chapter 6).

Einstein had strong opinions on a variety of nonscientific subjects, and he was not shy in expressing them. Thus, he strongly opposed any kind of nationalism, regarding it as source of needless conflict. In 1942 the writer of these lines was, at age sixteen, introduced to Einstein by the distinguished Hungarian mathematician Paul Erdos, with the words, "Professor Einstein, I would like you to meet a talented young Hungarian mathematician," to which Einstein responded, "Why Hungarian?"

Peter Lax
Professor Emeritus, Courant Institute
New York University

Preface

I have often been asked how I became involved with Albert Einstein's travel diaries and what inspired me to recast them in the form of this travelogue. Like most physicists, I have a long-standing fascination with Einstein, but the undertaking to create this book actually grew from my tenuous personal connection to him.

The story begins a few years ago, when I received a telephone call from a Mr. Andor Carius, an artist and scholar who I did not know at the time but who has since become a friend. As a student of Bengali culture and music, Andor was interested in two formal dialogues between Einstein and the celebrated Bengali artist, philosopher, poet, and composer, Rabindranath Tagore. These dialogues took place in Berlin in 1930, a time when there was great interest in Eastern art and philosophy among European intellectuals, who were inclined to perceive Tagore and Einstein as preeminent representatives of Eastern and Western culture, respectively. In their dialogues, the two men compared the Eastern and Western concepts of art, science, music, and religion, though, in the end, they found little common ground.[1]

In his research, Andor had discovered that the Einstein–Tagore meetings had been arranged by a certain Dr. Bruno Mendel and that one of the meetings had taken place in the splendid Wannsee villa where Dr. Mendel lived with his wife, Hertha; their three children; and Hertha's mother, the widowed Mrs. Toni Mendel.[2] The Mendel family had been friends of Albert and Elsa Einstein, who lived in Berlin from 1914 until 1933, and Toni Mendel, in particular, was a close friend and frequent companion of Einstein's. She and Einstein shared many interests, ranging from politics to music; they also shared leisurely outings in Einstein's magnificent sailboat, the *Tümmler*. Bruno Mendel, as an astute observer of the political scene, recognized the menace posed by Hitler very early and fled Germany with his family soon after the Nazis came to power. After working briefly in France and the Netherlands, he concluded that no European country was safe and brought his family across the Atlantic. They settled in Toronto, where Bruno worked as a medical researcher.

Andor had traced the subsequent migrations of the Mendel family, and although Bruno and Hertha were no longer alive, he had located their two surviving children, Ruth and Gerald, and learned that during the Second World War, their Toronto household had included two students who were refugees from Austria. These refugees had been released from a Canadian internment camp thanks to guarantees the Mendel parents made to the authorities. One of those young refugees was my friend Walter Kohn, and I was the other.

I had just turned fifteen when I made my escape from Vienna to England in April 1939, one year after Hitler annexed Austria. After the fall of France, there was widespread fear in England that a German invasion was imminent, and Churchill ordered the internment of all "enemy aliens" living near the coast. I was at the time employed as a dishwasher at a hotel in Brighton; I was duly arrested and interned, as was Kohn. Having attended the same high school in Vienna, we now met again in an internment camp on the Isle of Man. From there we were shipped to a series of camps in Canada, spending the longest time in a camp deep in the forests of New Brunswick where we worked as lumberjacks. The internees, mostly German and Austrian Jews, organized a school for the camp. Though small, it boasted a first-rate faculty, which prepared about a dozen students, including Kohn and me, for the McGill University matriculation exams, which we took while still interned. In time, the Canadian authorities recognized the harmless nature of their civilian internees and sanctioned the release of bona fide students—if they had Canadian sponsors.

While still in the New Brunswick camp, I received a letter from a Mrs. Hertha Mendel in Toronto, informing Kohn and me that she and her husband wished to sponsor us and were inviting us to live with their family following our release. They had chosen us after learning from a former inmate that we played recorder duets together in camp. As music lovers, the Mendels made the dubious assumption that if we were amateur musicians, we could not be all bad.

After many delays, I was able to reclaim my freedom and joined the Mendel family in their lovely home in Toronto. The house was furnished with possessions from their former home in Berlin, and it was permeated by their love of art and music. Although the war against Hitler was going badly at the time and cast a pall over everything, I could not have asked for a more congenial home or adoptive country. I spent a happy year and a half as a member of the Mendel family before joining the Canadian army.

The Mendel family's dowager mother was "Omama Toni," as we all called her. Her home in Oakville, just outside Toronto, was filled with books and works of art she had brought from Berlin. She retained a lively interest in politics and continued to correspond with her old friend Einstein, then living in Princeton. Unfortunately, after Omama Toni died, almost all of Einstein's letters to her were destroyed, but in one surviving letter (from 1948), he deplored the discouraging political scene of the time. In another, written in 1954, he thanked Toni for approving of his public denunciation of the McCarthy hearings and recalled their shared political and antiwar struggles of long ago in the days of the Weimar Republic.[3]

After the war, I completed my studies and eventually settled in the United States. I tried to stay in touch with the Mendel family, but soon we became widely dispersed. Bruno and Hertha moved back to their beloved Netherlands, where they lived in a small house by a canal near Bussum. My wife and I visited them there whenever we were in Europe. I recall an occasion in the 1960s when we sat with Hertha in her garden, savoring the pastoral calm while deploring the dispiriting political climate of the cold war. The hopes for an enlightened future that Einstein and the Mendels had shared so fervently in the days of the Weimar Republic remained unfulfilled.

Seventy years had passed since I was a member of the Mendel household when Andor asked me to dredge my memory for recollections of the Mendels. He, in turn, introduced me to the riches of the Einstein archive at Princeton University, where Einstein's travel diaries quickly caught my attention. They brought to life for me not only their extraordinary author but the troubled times in which they were written. They also brought back memories of Omama Toni, as for instance when Einstein, aboard a ship headed for New York, gave thanks to Toni for having supplied him with licorice candy for the journey. Who knew?

J. E.
New York
March 19, 2011

Acknowledgments

I am grateful to the Albert Einstein Archive, Princeton University Library, for allowing me access to the photocopies of Einstein's travel diaries, their transcripts, and to other Einstein material.[1] I am equally indebted to the Einstein Archive, Hebrew University of Jerusalem, for its kind permission to include short quotations from the travel diaries in my narrative.

I thank the staff of the Bodleian Library, University of Oxford, for making the Deneke material used in chapters 6 and 8 available to me; Jenifer Glynn for allowing me to quote from her mother's letter; and Paul Kent for sharing with me his knowledge of Einstein in Oxford. Many thanks, also, to Freeman Dyson for telling me about Einstein's visit to Winchester College.

I am most grateful to Alice Calaprice for being my constant guide in the world of book publishing and the world of Einstein biographical research. I thank Andor Carius for his expertise, particularly regarding the Einstein-Tagore dialogues.

The unstinting help and advice of Linda Regan—my editor—and of other members of the staff of Prometheus Books was much appreciated. Thanks to Will DeRooy for his careful and intelligent work in copyediting the manuscript.

I thank Walter Gratzer warmly for his critical reading of the entire manuscript and for sharing with me his inexhaustible store of knowledge of the history of science and scientists.

Finally, I am grateful to my wife, Styra Avins, for being accommodating toward my long preoccupation with Einstein and for her counsel in musical matters; I thank her, my daughter Alison, and my son Simon, for their loving support and skillful copyediting advice.

Introduction

I hope that this little book will conjure up for the reader Albert Einstein's world as he experienced it from 1922 to 1933 and, specifically, in the course of his far-flung journeys. Einstein was a prodigious traveler during those years, and whenever he was "on the road," it was his custom to keep a travel diary, a *Reisetagebuch*. In these journals—the only substantial diaries he ever kept—Einstein chronicled his daily activities and his observations on current events and personalities, but he also included his musings on topics ranging from physics and politics to art and music. These almost daily entries, made in his neat handwriting, provide an evocative glimpse into the ways in which he spent his waking hours and how he perceived his world.

At the time of these diaries, Einstein was in his forties. Although his most important work was behind him, he remained deeply engaged with contemporary science and with his ultimately futile search for a unified field theory. He also took an active part in the political and social life of Berlin, his home since 1914. His fellow scientists had celebrated Einstein since 1905, his annus mirabilis, when he burst like a meteor onto the physics scene; but it was not until 1919 that he became a worldwide celebrity on a scale previously unknown. His sudden fame came about not on account of his numerous contributions to physics, but as a result of a widely publicized finding that confirmed a critical prediction of relativity theory: the deflection of light by the sun's gravitational field, as measured by Arthur Eddington during the 1919 solar eclipse in Brazil.

Einstein was, by nature, a private person who, left to his own devices, preferred a bohemian lifestyle, a dress code with little use for socks or tuxedos, and a hairdo of his own design. But he was, alas, saddled with a powerful social conscience, and when fame was thrust upon him, he did not withdraw into himself but made use of his prominence to promote his favorite humanitarian causes. And, for the most part, he was willing to pay the price of shaking hundreds of hands at receptions, being pursued by reporters and photographers, and listening to innumerable accolades that meant nothing to him.

Einstein's travels all took place during the slow decline of the Weimar Republic, Germany's brave experiment in democracy, and the concomitant rise of Hitler. But the period is also noteworthy for the vibrant art, architecture, literature, film, and music that it produced—as well as far-reaching discoveries in physics and astronomy.

At the beginning of Einstein's travel decade, Europe had barely recovered from the trauma of the First World War. Germany's old political order had been overthrown along with the kaiser, but the republic that took its place won only grudging support from the public and was constantly undermined by the Far Right and the Far Left. Internationally, intense bitterness persisted on both sides of the recent conflict: for example, German athletes were excluded from the 1922 Olympic Games, and the last German prisoners of war in France were repatriated only that year. At the end of Einstein's travels, the Weimar Republic was in its death throes and Hitler was lurking in the wings.

Amid the turbulent politics of the 1920s, Einstein's immense prominence, together with his Jewish origins and his devotion to pacifist and humanitarian causes, made him a highly controversial figure in Germany. Einstein embarked on the longest of his journeys, a five-month tour of the Far and Near East, at a time when both his politics and his physics were under virulent attack in Germany, and he was in danger of assassination. In 1921 he was invited to lecture in Japan, and the lecture tour offered a chance not only to visit new and exotic places, but also to escape the turmoil in Berlin and to earn a sizable hard-currency fee. The voyage to and from Japan by ship left Einstein with an abiding fondness for this mode of transport: oceangoing vessels provided the tranquility he treasured, away from reporters and photographers.

The travel diaries Einstein kept were meant solely for his own use. It is therefore hardly surprising that his comments about people and events are candid, and that a few seem racially insensitive and even offensive; these must be seen in the context of the times. Among the many striking aspects of Einstein revealed in the diaries is his great devotion to music and to music-making—he hardly ever left home without his violin. Finding chamber-music partners, whether at sea or on land, was among his highest priorities.

While this account of Einstein's voyages leans heavily on the travel diaries, they have been explained and supplemented with material culled from many other sources, including contemporary newspapers. It should be understood, however, that wherever the narrative uses phrases such as *he thought, he was surprised by, he concluded, he wondered*, or *he mused*, these are always based on explicit comments in his diary. Similarly, when the narrative mentions that the night sky was misty, or the sea choppy, it should be understood that such information comes from Einstein. Direct (translated) quotations from the diaries always appear between *single* quotation marks (thus: 'It was really lovely in Leiden.') The occasional, interjected comments by the narrator are readily discernible as being clearly distinct from Einstein's.

In retracing Einstein's footsteps on his voyages, it is, of course, important to keep in mind their historical context and Einstein's personal history. Chapter 1 is therefore devoted to "setting the stage" for the chapters that follow, by summarizing the momentous historical events that preceded Einstein's travel decade, and by providing a synopsis of Einstein's life and work. Additional biographical, scientific, and historical information is to be found in the notes at the end of the book and in sources listed in the bibliography.

Timeline

1879   Albert Einstein born in Ulm, March 14.

1896   Einstein passes *Matura* examination in Aargau, enters ETH (Swiss Federal Institute of Technology).

1900   Einstein graduates from ETH but fails to obtain an assistantship.

1903   Employed at patent office, Bern, Einstein marries Mileva Marič.

1905   Einstein's annus mirabilis. Special theory of relativity published.

1909   Professorship, University of Zurich; Prague (1911); ETH (1912).

1914   Einstein moves to Berlin, separates from Mileva. WWI breaks out.

1915   General theory of relativity published.

1918   WWI ends. Wilhelm II abdicates. Birth of Weimar Republic.

1919   Einstein divorces Mileva; marries Elsa Löwenthal. Becomes an international celebrity.

1921   First American tour, with Chaim Weizmann. Einstein and Elsa arrive in New York, April 2, on TSS *Rotterdam*. Depart New York for home, May 30, on SS *Celtic*.

1922   Einstein and Elsa depart Marseille, October 7, on SS *Kitano Maru*. They visit Colombo, Sri Lanka, Singapore, Hong Kong, and Shanghai and extensively tour Japan. Depart Japan, December 29, on SS *Haruna Maru*. Arrive in Shanghai, December 31.

1923   Einstein and Elsa depart Shanghai, January 2, on SS *Haruna Maru*. Stop in Hong Kong, Singapore, Malacca, Penong, and Colombo. Disembark in Port Said, February 1.

Visit Palestine (Jerusalem, Tel Aviv, Haifa), February 2–14. Depart Port Said, February 16, on SS *Oranje*. Arrive in Toulon, February 21. Tour Barcelona, Madrid, and Saragossa, then head for home, March 14.

1925   Einstein departs Hamburg, March 6, on SS *Cap Polonio*. Stops in Boulogne and Bilbao. Arrives in Rio de Janeiro, March 21. Over next two months visits Buenos Aires, Córdoba, Montevideo, and Rio de Janeiro. Departs Rio for home, May 12, on SS *Cap Norte*.

1930   Einstein, Elsa, Dukas, and Mayer depart Antwerp, December 2, on SS *Belgenland*. Stay in New York, December 11–15, before traveling to Pasadena by way of Havana, the Panama Canal, and San Diego. Stay in Pasadena, December 31–February 28, 1931, with an excursion to Palm Springs.

1931   Einstein and Elsa arrive in New York by train, March 4, and depart for home, same day, on SS *Deutschland*, arriving in Hamburg on March 15.

Einstein departs Hamburg, April 30, on SS *Albert Ballin*. Arrives in Southampton, May 1; spends four weeks at Oxford. Departs Southampton for home, May 28, on SS *Hamburg*.

Einstein and Elsa depart Antwerp, December 2, on MS *Portland*. With stops in Panama, Honduras, El Salvador, and Los Angeles, arrive in Pasadena, December 31, and stay through March 1932, with another Palm Springs interlude.

1932   Einstein and Elsa depart Los Angeles, March 4, on MS *San Francisco*. Arrive in Hamburg, March 29.

Arrive in Cambridge around April 20. Visit Oxford, April 29–ca. June 1.

Einstein and Elsa depart Bremen, December 10, on MS *Oakland*. Arrive in Antwerp, December 11. Depart Antwerp, December 14.

1933   Einstein and Elsa arrive in Los Angeles, January 9. Stay in Pasadena through March 11, with another desert excursion. Travel by train to Chicago and New York. Depart New York, March 18, on SS *Belgenland*. Arrive in Antwerp, March 28.

Einstein and Elsa stay in Le Coq sur Mer, Belgium, ca. April 1–September 8. Einstein departs for Cromer, Norfolk, September 8. Departs Southampton, October 7, on SS *Westernland*. Arrives in New York, then Princeton, October 17.

1935   Einstein, Elsa, stepdaughter Margot, her husband, and Dukas depart New York, May 25, on SS *Queen of Bermuda*. Arrive in Hamilton, Bermuda, May 27. Depart Hamilton, June 1, on *Queen of Bermuda*. Return to New York, June 3.

1936   Death of Elsa Einstein.

1940   Einstein becomes US citizen.

1955   Death of Albert Einstein, April 16.

1.

Setting the Stage

In 1912, while Einstein occupied the chair for theoretical physics at Prague's German University, he visited Berlin for the first time. At that stage, he had already won wide recognition among physicists for his work in quantum physics and for his special theory of relativity, but he was by no means the celebrity he was to become in the 1920s. His weeklong stay in Berlin gave him an opportunity to meet with Max Planck, whom he greatly admired for having introduced quanta into physics, and with other renowned scientists then working in Berlin, Europe's preeminent center of physics research.[1] But Einstein also used his visit for an affectionate reunion with his cousin Elsa Löwenthal, whom he had known since childhood. She had recently divorced and was now living in Berlin with her two teenage daughters.

Two years later, Einstein was offered a munificent academic position in Berlin, truly an offer he could not—and did not—refuse. He consequently took up residence in the Prussian capital in 1914, arriving with his wife, Mileva, and their two young sons. But the marriage ended in acrimony soon after, and a few months later, the First World War broke out. Einstein's divorce did not become final until 1919, and when it did, he married Elsa. The couple established their household in Berlin, and the city remained their home until the Nazis came to power in January 1933. It was therefore as a Berliner that Einstein experienced the bleak years of World War I, their violent aftermath, and the bracing and turbulent years of the Weimar Republic, until its demise.

The present chapter sets the stage for Einstein's travel decade by reviewing the historical events that preceded it and by providing an outline of Einstein's earlier life and work.

BACKGROUND: WILHELMINE BERLIN AND THE RUSH TO WAR

At the time of Einstein's visit to Berlin in 1912, the German Reich was just forty years of age. It had been assembled from numerous German principalities, most of them herded together under the dynamic leadership of Prussia, the Reich's most powerful member state. The emergence of a united Germany in 1871 represented the fruits of three victorious wars that Prussia waged against Denmark, Austria-Hungary, and France and also of the tireless diplomatic efforts of Prussia's prime minister, Count Otto von Bismarck. The newly created German state was nominally a constitutional monarchy, governed by a council, with limited input from an elected parliament, the Reichstag. In practice, however, most of the power was vested in the person of the kaiser, who was also king of Prussia, for the constitution gave him the authority to dismiss the governing council, as well as the Reichstag. Bismarck remained the Reich's guiding spirit from its inception until 1890, when he fell out with the newly anointed Wilhelm II. The new kaiser was by all accounts a vain and insecure young man who was not prepared to be overshadowed by his principal minister. The break came over a dispute regarding social policy, but its most profound effect was in foreign policy, for Wilhelm dismantled Bismarck's policy of negotiating interlocking alliances with the other great European powers as the means of ensuring Germany's security.

Wilhelm II put a more aggressive policy in place: he declined to renew the treaty with Russia that Bismarck had negotiated, and he was vociferous in his demands for "a place in the sun" for Germany alongside the European empires that were already in possession of the most desirable colonial real estate. Above all, Wilhelm was jealous of the global power exercised by Great Britain, which was ruled by his grandmother, Queen Victoria. He felt that his English relatives did not pay him the respect due to an emperor, and he was particularly envious of his uncle Albert, the Prince of Wales. He alienated "Uncle Bertie" by his boisterous behavior and poor sportsmanship during the Cowes regatta, that extravagant annual gathering that drew the cream of the European nobility to the Isle of Wight. It was in yacht racing that Wilhelm and Albert acted out their rivalry during the 1890s. Wilhelm even established his own grand regatta in Kiel, in an attempt to rival the Cowes regatta, but it was not a success. Queen Victoria tried hard to mend relations between the two ruling houses. She even offered her difficult grandson Wilhelm the title of Honorary Admiral of the Fleet, but to no avail. The breach between the two dynasties was further exacerbated when Wilhelm meddled in the Boer War, sending Paul Kruger, the president of the short-lived Boer Republic, a congratulatory telegram that hinted at Germany's support against Britain.[2]

Perceiving Britain as a greater threat to Germany than Russia and France, Wilhelm ordered the rapid expansion of his Imperial Navy so that it might challenge the supremacy of the Royal Navy, Prince Albert's pride and joy. Admiral Alfred von Tirpitz was charged with building a powerful new battle fleet—a provocation that Britain, with her many far-flung possessions, could hardly ignore. Britain responded by commissioning even faster, better-armed, and better-armored battleships, such as HMS *Dreadnought,* and Germany responded in kind. As the arms race between the two navies escalated, contemporary observers perceived it as a contest to decide who would rule the world.

On land, Wilhelm appointed Alfred von Schlieffen as army commander, to succeed Helmuth von Moltke, the celebrated hero of the Franco-Prussian War. Schlieffen's strategic plan for crushing France called for German armies to thrust west through Belgium before turning south, toward Paris, while employing a blocking force in the east to contain Russia.[3] The plan recognized that a quick victory in the west was essential, since Germany could not win a protracted war. The Schlieffen plan was the strategic basis of the Triple Alliance of Germany, Austria-Hungary, and Italy, which was arrayed against the *Entente* powers: Britain, France, and Russia. Historians have offered many reasons for the nations' headlong plunge into war, but it cannot be denied that Wilhelm's personality flaws and his espousal of militaristic nationalism played a significant role. Einstein was, of course, thoroughly familiar with the history of Wilhelmine Germany, and nothing could have been more repugnant to his instinctive pacifism than the kaiser's policies.

A word, finally, about the internal politics of Wilhelmine Germany. The rapid industrialization and urbanization during the second half of the nineteenth century led to enormous shifts in the German population and to growing influence of the working classes and the Social Democratic Party (SPD). Bismarck tried to limit the political power of the socialist parties, and at one time actually outlawed them, but in the wake of the 1912 general election, the SPD became the largest party in the Reichstag. Since that was Germany's last general election until after the war, the SPD had a powerful voice in the creation of the Weimar Republic.

BEFORE BERLIN (1879–1914)

Soon after Einstein settled in Berlin, the war that many had seen as inevitable became a stark reality. The exultation with which the German public welcomed war appalled Einstein, particularly because this war euphoria infected many of his academic colleagues. Vienna, London, and Paris greeted the outbreak of war with similar enthusiasm. Einstein's dismay must be seen in light of his lifelong loathing of militarism and of all herdlike behavior of people. Many years later, he wrote that any man who enjoys marching in formation earned his loathing, for that man had surely obtained his large brain in error: "Heroism upon command, senseless brutality and wearisome patriotism, with what fervor I despised them, and how base and despicable war seems to me. I would rather let myself be cut to pieces than participate in such evil doings!"[4] Only when Hitler's rise to power was complete did Einstein modify his pacifist principles.

It was his abhorrence of the Prussian-style militarism that prompted the fifteen-year-old Einstein to leave his native Germany in 1894 and to relinquish his Württembergian and German citizenship two years later. He remained stateless until he won his dearly treasured Swiss citizenship four years thereafter.

How, then, was it possible that twenty years later, Einstein found himself in the employ of the Prussian state and living in its capital? To understand this paradox, it is necessary to recall Einstein's life and work before he took up residence in Berlin.[5]

*   *   *

Einstein's ancestors on both his parents' sides belonged to the Swabian Jewry that had long resided in the many small towns of Swabia, a region in southern Germany that is now part of the state of Württemberg. His father's family had lived in the little town of Buchau for generations, in times when Jews were severely restricted in choosing their residence and occupation. During the course of the gradual emancipation of German Jews in the wake of the Napoleonic era, many restrictions gradually disappeared, and by the second half of the nineteenth century, enterprising Jews were able to enter new trades and to migrate from their native small towns to larger cities. Several of the Buchau Einsteins migrated to the nearby ancient city of Ulm; Einstein's father, Hermann, was among them. In 1876, Hermann married Pauline Koch, who had a similar Jewish-Swabian background, although hers was of a worldlier, well-to-do family that operated a successful grain business.

On March 14, 1879, Hermann's and Pauline's son, Albert, was born into this closely knit yet geographically dispersed family, a circumstance that, many years later, explains how Einstein could call on relatives in several far-flung places that he visited on his travels. Albert was described as an amiable child who began to talk very late but who, once he did, would silently construct complete sentences before articulating them. As a child, Albert was given to occasional violent temper tantrums; he did not enjoy playing with other children, preferring his own company. It was said that when he was among children, he conveyed an aura of isolation—an aura he would retain all his life.

Shortly after Albert was born, Hermann joined his youngest brother, Jakob, in a business venture that obliged the family to move to Munich, and soon after that the family was augmented by the birth of Albert's sister, Maja. But even though the family had left Ulm, they never lost their soft Swabian dialect or the Swabian fondness for diminutives: to his family, Einstein would always remain their "Albertle." When Einstein was six, he entered the Catholic public school, where he was the only Jew in his class and was exposed to anti-Semitic taunting by his schoolmates. He was a very good, if not an exceptional, student, both in this school and later, in the Luitpold Gymnasium in Munich. He excelled in mathematics, but he chafed under the school's strict discipline, and he resented having to study a subject merely to pass an examination. At the Luitpold Gymnasium, the teachers laid the greatest stress on Latin and Greek and paid scant attention to mathematics and science, subjects that Einstein studied on his own.

Einstein's mother, Pauline, played the piano, and it was she who introduced young Albert to music. He received his first violin lessons at age six, but he rebelled against his teachers' "mechanical" approach. At age thirteen, when he became acquainted with Mozart's violin sonatas, his passion for music was aroused, and he soon became a largely self-taught but proficient amateur violinist. When Einstein was in his twenties, living in Bern and Zurich, he was in great demand as a chamber-music player. He remained deeply devoted to music for the rest of his life.

Like most members of the largely assimilated German Jewry, Einstein's parents acknowledged their Jewish heritage freely but paid little heed to the observance of religious practices. As a result, Albert received his first religious instruction at the Luitpold Gymnasium and passed through a brief period of religious fervor. While it lasted, he refused to eat pork at home, and he composed hymns to the greater glory of God. But shortly before his scheduled bar mitzvah, this religiosity came to an abrupt halt when he had his first encounter with science and he disavowed formal religion. Many years later, Einstein recalled that the popular science books he read at the time had convinced him that the stories in the Bible could not be true, and this had made a devastating impression on him. He became a fanatical free-thinker with a "skeptical attitude towards all beliefs that happen to be prevalent in the current social environment"—an attitude he retained all his life. [6]

The science books that affected Einstein so profoundly were presented to him by Max Talmud, a medical student in Munich who was the Einsteins' dinner guest every Thursday evening—possibly in approximate deference to the Jewish tradition of inviting an impoverished Bible scholar to the weekly Sabbath evening meal. Talmud also presented Albert with a Euclidean geometry book that affected the boy deeply: "Here were assertions . . . that can be proved with such certainty that any doubt seemed out of the question. This clarity and certainty made an indescribable impression on me."[7]

Jakob Einstein, Hermann's business partner, was a graduate engineer, the only one of five siblings to attend an institution of higher learning. He took charge of the technical aspects of their engineering firm, while Hermann was the business manager. The firm manufactured dynamos, arc-lamps, and other electrical equipment, and it benefited from the boom in the embryonic electrical industry at the end of the nineteenth century. The enterprise prospered initially, but when the brothers' bid for the construction of Munich's lighting system lost out to much larger electrical engineering concerns, the firm was forced out of business. In 1894, the two brothers entered into another engineering venture, this time in Italy, and the two families moved to Milan. Albert was left behind in Munich, however, to live with a distant relative, so that he could complete his studies at the gymnasium and earn his *Abitur* (high school graduation) certificate—a sine qua non for any presentable member of the German middle class.

Einstein missed his family dearly, and he chafed under the school's rigid teaching methods, but he stuck it out at the gymnasium for half a year. Determined to escape, he managed to obtain a letter from a physician—Max Talmud's older brother—stating that he was undergoing a nervous breakdown.

The letter called on the school to release him for six months so that he could recuperate in the care of his parents. The school obliged, and in December 1894, Einstein appeared unannounced at his parents' doorstep in Milan. He assuaged their serious misgivings by promising that he would study on his own all summer and would then apply for admission to the prestigious Polytechnikum in Zurich (now the Swiss Federal Institute of Technology, or ETH).

Einstein enjoyed his Italian summer thoroughly, but he failed to pass the ETH entrance examination in the fall. His examiners did, however, recognize that they were dealing with a child prodigy, and they urged Einstein to attend the cantonal high school in Aarau for one year in order to catch up in his two weakest subjects, French and chemistry, before applying again. Einstein followed this excellent advice, and his school experience in Aarau contrasted sharply with that at the Luitpold Gymnasium. He was fortunate in being able to lodge in the home of Jost Winteler, a teacher at the cantonal school, with whom Einstein developed a close and long-lasting friendship, as he did with all members of the Winteler family. They made Einstein feel like one of them, and Jost Winteler's liberal political views probably played a role in Einstein's decision to renounce his German citizenship. Several years later, Einstein's sister, Maja, married a son of the Wintelers, making Einstein truly a member of the family; and later one of the Winteler daughters married Michele Besso, one of Einstein's closest and oldest friends.

Einstein passed his final high school examination, the *Maturitätsprüfung*, in 1896 and was admitted to the Polytechnikum. As a student, he supported himself on a modest monthly stipend of one hundred franks, provided by the Kochs—relatives on his mother's side—and he supplemented it with earnings from tutoring.

The 1890s were an exciting era in science in which X-rays, quanta, and electrons were discovered, and physics was emerging from its classical phase. Einstein studied the works of the exponents of the new physics—Helmholtz, Kirchoff, Boltzmann, Maxwell, and Hertz—but found the lectures of his professors at the ETH uninspiring and often skipped them. He held the firm belief that primitive mathematics was sufficient for formulating physical laws, and he did not attend the mathematics lectures of Hermann Minkowski. Ironically, it was Minkowski who, a few years later, gave relativity the elegant formulation familiar to physicists to this day, by expressing it in terms of a four-dimensional space-time continuum. It was Professor Weber's physics laboratory that held the greatest attraction for Einstein, but

Einstein's intellectual independence and contempt for convention did not endear him to Weber or any other of his professors at the ETH. Nevertheless, he obtained excellent examination grades thanks to the meticulous lecture notes kept by his conscientious fellow student, Marcel Grossmann, who loaned them to Einstein. The two men were to remain lifelong friends.

The friendship Einstein developed with Mileva Marič, a student at Zurich University, was even more fateful. She had entered the university at a time when it was the only German-language university to accept women as students. She came from a Greek Orthodox family in Vojvodina, then part of Austria-Hungary, and her native tongue was Serbian. To the surprise of his friends and to the dismay of his parents, Einstein became deeply infatuated with Mileva and eventually married her, over the strenuous objections of his family.

By 1900, the year he graduated from the Polytechnikum, Einstein was determined to pursue a career in physics. His high grades led him to hope and expect that he would be offered an assistantship by one of his professors, the usual point of entry to academia. When no offers came his way, he applied for assistantships at several German and Dutch universities, but again, without success. His attempts to find a high school teaching position were equally fruitless, and he blamed all of these failures on Professor Weber, with whom he had fallen out during his student days. As a result, Einstein found himself without an income and separated from Mileva, who had gone to stay with her family in Novi Sad, where she gave birth to his baby daughter. The girl was quietly given up for adoption.

With his fortunes at low ebb, Einstein's friend Marcel Grossmann came to his rescue once again. He introduced Einstein to officials of the Swiss patent office in Bern, who were favorably impressed and offered him a position as a patent examiner. Now that Einstein finally had a steady income, he had Mileva come to join him in Bern, where the two were married on January 6, 1903. The couple established a modest household together, and a year later their first son, Hans Albert, was born.

His work at the patent office suited Einstein perfectly. It was varied and technically challenging, and it still left him plenty of time to pursue his scientific investigations. The job also left Einstein with an abiding interest in inventions and patents. Many years later, well after he had achieved fame and recognition as a physicist, he worked on inventions, owned patents of his own, and earned royalties from them.

During his years in the patent office, Einstein was able to acquire a comprehensive knowledge of the current state of physics. He was a regular reader of the *Annalen der Physik*, the most important physics journal of the time, and became thoroughly acquainted with Planck's quantum hypothesis as well as Lennard's experiments on the photoelectric effect, Maxwell's electromagnetic field theory, Boltzmann's statistical mechanics, and the current theories of "the ether"—which was thought to fill all space and to be necessary for light propagation. In 1901, the *Annalen* published Einstein's first article, in which he deduced the nature and strength of molecular forces from the measured surface tension of a liquid—this at a time when many scientists had not accepted the reality of atoms and molecules, let alone quanta. The topics of Einstein's early papers ranged from statistical thermodynamics to molecular physics, and they impressed the editors of the *Annalen* sufficiently for them to use him as a manuscript reviewer, which gave him even greater access to the current physics literature.

Then came Einstein's annus mirabilis (1905), the year in which he published the four papers that are at the foundation of twentieth-century physics: (1) the special theory of relativity, including the equivalence of mass and energy ($E = mc^2$); (2) a molecular theory of liquids, which allowed him to derive the size of molecules from diffusion and viscosity data; (3) a statistical theory of heat that explained the random (Brownian) motion of small particles suspended in liquid; and, finally, (4) a corpuscular theory of light that provided a detailed explanation of the photoelectric effect. Taken together, these papers firmly established the existence of molecules and light quanta (photons), and they banished the ether hypothesis once and for all.

The equivalence of mass and energy, demonstrated in the first of these papers, was a harbinger of nuclear physics, and it elucidated the energy source of the sun and the stars. Einstein submitted the second paper to the University of Zurich as his PhD dissertation, and upon its acceptance, he finally reached the threshold of the academic career that had eluded him five years earlier. The contention of the last paper, that light is quantized (i.e., that it resembles particles, known as photons) was accepted only slowly by the physics community, since it appeared to be at variance with Maxwell's electromagnetic field equations and with many phenomena (diffraction, interference) that demonstrate the wave nature of light. Even eminent physicists, including Michelson and Planck, remained skeptical for a long time, but seventeen years later, after many experiments had verified the quantum hypothesis of light, Einstein was awarded the Nobel Prize for this work.

Throughout this period, Einstein continued to perform his daily duties as Patent Examiner Class 3, never stopping to catch his breath. In the years that followed, he published papers of similarly fundamental importance that dealt with the nature of radiation, molecular dynamics, statistical thermodynamics, and special relativity. He also developed a quantum theory of solids, in which the vibrational energy of the constituent atoms is quantized, a bold step that initially left even supporters, who now included Planck, unconvinced. But Walther Nernst's low-temperature experiments verified Einstein's predictions for the specific heat of solids and turned Nernst into an early champion of Einstein.

It had, by now, become evident to most physicists that Einstein's extraordinary scientific output was incompatible with his continued employment as a patent clerk. In 1909, after overcoming many academic, political, and bureaucratic obstacles, the University of Zurich offered him a professorship extraordinaire, created specially for him. He and Mileva abandoned their semibohemian life in Bern and moved to Zurich, where their expenses, but not their income, were considerably higher. When their second son, Eduard (Tetel), was born in the following year, the couple was obliged to take in lodgers to make ends meet. They were nevertheless happy to be back in the city they had come to love as students. Einstein had entered academia without passing through the customary apprenticeship as a lowly *Privatdozent* (assistant professor), and as a result he was unprepared for the arduous teaching load that went with his appointment: he was required to give seven lectures a week, and their preparation left him less time for research than he had enjoyed at the patent office.

Einstein was now widely recognized as a brilliant physicist in his prime. In 1909 he was invited to give a major lecture at a scientific meeting in Salzburg, his first opportunity to meet the great minds of contemporary physics, including Sommerfeld, Planck, and Born. Contrary to the expectations of the assembled scientists, Einstein did not speak about relativity; instead he spoke about the unsatisfactory state of quantum physics, which he considered the most serious challenge facing physics. Few in the audience agreed with that assessment or with his prediction that a new dynamics would have to emerge to deal with quanta—a prediction that was confirmed sixteen years later, with the discovery of quantum mechanics.[8] At the celebrated first Solvay Conference in 1911, which assembled the world's finest theoretical physicists, Einstein again cited the urgent need for a new mechanics, but he again found little support among his colleagues.

It was merely a matter of time before a full professorship at a major university would be offered to Einstein. In 1911, the chair for theoretical physics at Prague's German University (there was also a Czech university)—one of Europe's oldest universities—became vacant, and Emperor Franz Joseph approved his academic appointment. Einstein's salary was now substantially higher, but he and Mileva did not find life in Prague congenial—they missed their beloved Zurich. Einstein chafed under the weight of paperwork (the bureaucracy in the Habsburg monarchy notoriously cherished paperwork), and he complained in letters to his friends in Zurich that his students in Prague lacked interest in his field. He was also distressed by the mutual hostility between Prague's Czech-speaking and German-speaking citizens. His Swiss friends went to great lengths to persuade the government to offer him a professorship at the ETH, and when their efforts succeeded, the entire Einstein family was overjoyed. The irony of obtaining such a prestigious appointment at the same institution that, not many years earlier, had denied him a lowly assistantship was not lost on Einstein.

Einstein's comparative isolation in Prague had allowed him to return to a problem long on his mind, one that neither he nor others had been able to solve: while the 1905 version of relativity theory covered mechanics and electromagnetism, no one had succeeded in integrating gravitation into it. Einstein was, of course, aware that his theory was incomplete, or "special," because it applied only to systems that moved uniformly and linearly with respect to each other; it ignored accelerating systems. In what Einstein later described as the happiest thought of his life, he came to the realization that for a freely falling observer, the gravitational field ceases to exist. He concluded that gravity and acceleration were equivalent, since an observer in a closed laboratory had no way of distinguishing between them. Einstein postulated, furthermore, that special relativity's principle of equivalence—that within each system all physical laws, and the velocity of light, are identical—applies also to uniformly accelerating systems. But in implementing these two simple postulates, Einstein encountered considerable mathematical difficulties, which he was unable to overcome either in Zurich or in Prague.

That is why Einstein's move to the ETH in July 1912 could hardly have occurred at a more opportune time. As soon as he arrived in Zurich, he called on his old friend Marcel Grossmann, now a professor of mathematics at the ETH, and implored him to help solve his mathematical dilemma. Grossman came through once again, for he recognized that Einstein's postulates led to a space-time manifold with a non-Euclidian geometry, one that had been studied in great detail by the great Karl Friedrich Gauss and other mathematicians. Einstein eagerly followed this promising lead, but it nevertheless took four more years of hard work before he completed his *general* theory of relativity, arguably among the finest achievements of the human intellect.

When Einstein visited Berlin in 1912, the local academic luminaries were already hatching plans to entice him to their city, and in the following year their plans came to fruition. Einstein had barely settled at the ETH when Planck, Nernst, Fritz Haber,[9] and their colleagues at the academy and the University of Berlin presented Einstein with a bouquet of academic offerings designed to be irresistible. Their offer included membership in the Prussian Academy, a professorship at Berlin University, the directorship of his own physics institute,[10] a very generous salary, and what may have been the most compelling lure: no teaching duties at all.

At the ETH, Einstein had to cope with a massive teaching load that left him little time for finishing the relativity theory. Ridding himself of his onerous academic obligations was surely a reason to accept the offer, and so was the prospect of being associated with such esteemed colleagues in Berlin, not to mention being near to his cousin Elsa, with whom he had exchanged affectionate letters since their recent reunion. When Planck and Nernst traveled to Zurich and made their enticing offer to Einstein, its allure outweighed his aversion to things Prussian, and he accepted with alacrity.

IN BERLIN: WAR AND ITS AFTERMATH (1914–1922)

Einstein arrived in Berlin in April 1914 to take up his fourth academic post since leaving the patent office five years earlier. But this one he would retain for the next nineteen years. While he had overcome his misgivings concerning the move, the same was not true of Mileva, who was acutely apprehensive of the proximity to the Einstein family, who had never approved of her. Her fears were justified. Within two months, their marriage, which had been troubled for some time, fell apart for good. Einstein's friend Michele Besso came to Berlin to accompany Mileva and the two boys, aged ten and four, back to Zurich. There was an emotional farewell at the railway station, and Einstein returned home in tears—according to his newfound friend Fritz

Haber, who had accompanied him to the station. Mileva and the two boys were profoundly shocked by the separation; it took four years of bitter negotiations before a final divorce settlement was agreed on, during which time Einstein was unable to visit his sons.

It is remarkable that amid the turmoil surrounding his disintegrated marriage and under the gathering war clouds, Einstein was able to put the finishing touches to the general theory of relativity. Although he was now surrounded by several renowned scientists, few of them voiced support for the new theory when he presented it in his inaugural lecture before the Prussian Academy. In the absence of experimental evidence, Planck was outright skeptical. Among Einstein's few supporters was Erwin Freundlich, who planned to test the theory by measuring the deflection of light by the sun's gravitational field. Relativity theory predicted the magnitude of the deflection to be just barely observable, and observable only during a total solar eclipse, when the intense light from the sun is blocked by the moon. The first eclipse that lent itself for the experiment occurred in southern Russia in August 1914, and with Einstein's help, Freundlich raised sufficient funds to mount an expedition to Crimea. Unfortunately, the war broke out a few days before the eclipse, and Russia arrested Freundlich and the other expedition members with their equipment. They were exchanged for a group of Russian officers two months later, but the verification of general relativity had to wait until after the war.

In August 1914, Kaiser Wilhelm ordered that the Schlieffen plan be executed. German armies invaded Belgium before turning south and heading for Paris. A month later, advance units were close enough to Paris to see the glow of the city's lights, but French and British armies, and fierce Belgian resistance, forced the exhausted German units back to a line that was to change little in the four years of brutal trench warfare that followed. The quick victory envisaged in Schlieffen's plan had eluded Wilhelm, and the matériel advantage enjoyed by the Allies made the war's outcome inevitable.

As an instinctive pacifist, Einstein had always abhorred war. When it arrived, he became a political activist. He was especially distressed by the enthusiastic and unquestioning patriotism of his associates: Planck, whom Einstein much admired, exhorted his students to take up the sword against England, "the breeding ground of perfidy," while Nernst volunteered to serve as a courier at the front, using his own automobile. Fritz Haber placed the facilities of his Kaiser Wilhelm Institute at the disposal of the war ministry, and he worked on the practical problem of producing and deploying poison gas at the front. Ironically, Einstein's office was located in Haber's institute, so he must have been aware of Haber's war work. Still, his own productivity did not suffer during the war years. He not only completed the general theory of relativity but published fundamental papers in cosmology, thermodynamics, condensed matter physics, and quantum theory.

In the early weeks of the war, the invading German armies were accused of committing atrocities against Belgian civilians, and a consortium of eminent German intellectuals responded to these charges with a manifesto entitled "Appeal to the Cultured World." It claimed that Germany, "a cultured nation with the sacred legacy of Goethe and Beethoven," was not responsible for the war—that it was innocent of the charges of brutality. It accused the Allies, in turn, of driving "Mongols and negroes against the white race." The manifesto was signed by ninety-three illustrious German artists, writers, and scholars, among them Planck, Nernst, and Haber. This patent betrayal of intellectual honesty shocked Einstein. He and the like-minded physician Georg Nicolai drafted a restrained countermanifesto, but they were able to attract only two additional signatories. For the remainder of the war, Einstein could do little more than rail against the war madness surrounding him in letters to his Swiss friends and to the French writer Romain Rolland, a committed pacifist like himself.

In 1916, Wilhelm appointed Generals Hindenburg and Ludendorff as supreme commanders of the German war effort, and they soon exerted a growing influence in the political arena, as well. Ludendorff was an advocate of total war, and he ordered unrestricted submarine warfare, even though this policy was likely to draw the United States into the conflict—as it did. In the east, the German armies had been victorious, and Ludendorff negotiated the treaty of Brest-Litovsk with the Russian (now Bolshevik) government, which ceded to Germany vast territorial gains stretching from the Baltic to the Ukraine. Equally important, the treaty allowed Ludendorff to shift additional resources to the western front.

By 1918, the British naval blockade was having a telling effect, and the daily bread ration of German civilians was cut to five ounces. Ludendorff placed his last hope for victory on a massive spring offensive in the west. The offensive was initially successful, with German units pushing to within a hundred miles of Paris. But then the Allied counteroffensive pushed the front back to its starting line. At the battle of Amiens, hundreds of Allied tanks broke through the German lines. Ludendorff recognized that the situation was hopeless and informed the kaiser that the war was lost. The armistice that was agreed upon called for the removal of the kaiser and the democratization of Germany ("regime change," in modern parlance), as well as the withdrawal of German armies from all conquered territories in both the east and the west. In a whirlwind sequence of events during the following months, the kaiser fled the country, German workers and soldiers rebelled and hoisted a red flag over the imperial palace in Berlin, sailors mutinied in Kiel, and street fighting broke out in several cities. Out of this turmoil, the Weimar Republic was born, its existence under constant threat from its enemies on the left and the right.

Meanwhile, Ludendorff contrived to have the armistice negotiations conducted not by the military, but by the newly formed social democratic government, headed by Friedrich Ebert. Since the public had been led to believe until the very end that victory was within Germany's grasp, and German armies stood deep inside France and Russia, this sudden military collapse shocked the civilian population profoundly. Many gave credence to the so-called *Dolchstosslegende*, or "dagger thrust legend," the myth that the army had not been defeated but had been stabbed in the back by the politicians. This claim was to become a rallying cry of the Nazis.

Einstein's divorce settlement in February 1919 contained the provision that if Einstein were awarded the Nobel Prize, the prize money would go to Mileva. Four months later, Einstein married Elsa and found himself the head of a household that included Elsa's two daughters, aged twenty-two and twenty.

A few days before the marriage, Arthur Eddington announced at a meeting of the Royal Society in London that he had measured the deflection of starlight by the sun's gravitational field. The observed deflection was in agreement with what general relativity theory predicted, and Einstein became a world celebrity almost overnight.

FIRST VISIT TO AMERICA (1921)

It is hardly surprising that Einstein welcomed the demise of the old imperial order and became an enthusiastic supporter of Germany's young democracy. Amid rising anti-Semitism, Einstein was attacked virulently for his pacifist sentiments and even for his scientific work, and his public physics lectures were often picketed and disrupted. In this atmosphere, he was persuaded that in the future, Zionism offered the best hope for European Jews—though without hiding his own indifference to religious Judaism or deeming himself a Zionist. In 1921, Chaim Weizmann, himself a scientist and the newly elected president of the Zionist World Organization, asked Einstein to join him on a fundraising tour of the United States.[11] Einstein agreed, in spite of his reservations. Because he was aware of Einstein's ambivalent attitude toward Zionism, Weizmann did not invite him to speak at their joint public appearances. He merely used him and his enormous prominence to attract audiences and donors for the Zionist cause—and, specifically, for the planned Hebrew University in Jerusalem. As Einstein later wrote in a letter to his friend Michele Besso, "I had to let myself be exhibited like a prize ox."[12]

Elsa accompanied her husband on this trip, their first to America, and while Einstein did not keep a travel diary, it is evident from the voluminous newspaper coverage that he and Weizmann received tumultuous welcomes wherever they went. On their arrival in New York, they were welcomed by Zionist functionaries. A corps of voracious journalists questioned Einstein closely about his theory, until he finally quipped, "I hope I have passed your examination." He then submitted to being exhaustively photographed for half an hour before making his escape. Thereupon, Elsa took on the reporters and explained to them that her husband disliked being a public exhibit and that he would rather work and play his violin or take a walk in the woods. Asked whether she understood his theory, Elsa answered with a laugh: "Oh, no, although he has explained it to me so many times; I understand it in a general way, but in its details it is too much for a woman to grasp. But it is not necessary for my happiness." In another newspaper interview, the guileless Elsa told reporters that when her husband was engaged in some problem "there was no day and no night" for him, but at other times, he went for weeks without doing anything in particular other than dream and play his violin. Whenever he was weary in the midst of his work, he went to the piano or picked up his violin and soothed his mind with music. She added that he was particularly fond of Mozart but that Brahms was another of his favorites.[13]

Einstein was unprepared for the tumult he caused in America. Speaking at the National Academy of Sciences, he explained that when a man, after long years of searching, chances upon an idea that reveals something of the beauty of this mysterious universe, he should not, for that reason, be personally celebrated, for he was already sufficiently rewarded by the experience of seeking and finding. In science, one's work is, moreover, so bound up with that of one's scientific predecessors that one's own achievement almost seems to be the product of theirs.[14]

On their journey home, Einstein and Elsa stopped for a few days in London, where he was to receive the Gold Medal that the Royal Astronomical Society had awarded him. The ceremony, however, was blocked by members who still felt hostile toward Germans after the war. Einstein laid flowers on Isaac Newton's tomb in Westminster Abbey and lectured on relativity at King's College, where the initial frosty welcome gave way to a standing ovation following his lecture. He had become a skilled and effective ambassador of goodwill toward the scientists of the former enemy nations, and he pursued the goal of better understanding between them.

In Germany, however, Einstein's pacifist activities abroad were viewed with suspicion. Relativity theory continued to be attacked by rightist zealots, who were led by two Nobel laureates, Philipp Lenard and Johannes Stark. Their arguments were at first philosophical in nature but became increasingly chauvinistic and anti-Semitic.[15] In 1920, after Einstein and Lenard became embroiled in polemical exchanges in the press and at public meetings, Einstein toyed with the idea of leaving Germany, but many voices urged him to remain.

A DINNER AT THE EINSTEINS'

The diplomat and art collector Count Harry Kessler was the son of a prominent German banker with close ties to Prussian royalty. Kessler's considerable wealth and social connections enabled him to move freely in all circles of society. He was an important patron of the arts and artists, but today he is remembered chiefly as an acute observer and dedicated chronicler of the artistic, social, and political life in postwar Berlin. With the fall of Imperial Germany, Kessler became an avid supporter of the Weimar Republic, and he was widely known as the Red Count. His political philosophy had much in common with Einstein's, and, having met at various social and political functions, the two became friends. The diaries Kessler kept from 1918 until his death in 1937 contain many perceptive accounts of the events and personalities of the Weimar period that capture the temper of the times.

Kessler's diary entry for March 20, 1922,[16] depicts the mood of Einstein and Elsa, shortly before they set out on their epic journey to the Far East. On that day, Kessler breakfasted at the foreign ministry as the guest of Walther Rathenau, a proud, brilliant, and wealthy businessman who had recently been appointed foreign minister of the Weimar government. Rathenau and Kessler were friends and political allies. Rathenau was continually subjected to vicious attacks from the extreme right on account of his Jewish roots, and over breakfast he complained bitterly to Kessler of the heavy burden of his office. He said that the worst aspect of his post was the malevolent opposition he encountered inside Germany, and the personal affronts he had to endure in silence. He also spoke of how difficult it was to convince his French negotiating partners that Germany would uphold the spirit of disarmament, even while an entire generation of young Germans was sliding toward the worst sort of reaction. Rathenau mentioned receiving threatening letters every day and that the police had given him serious warnings. With that, Rathenau pulled a Browning pistol from his pocket and told Kessler that it had come to this: he no longer left his home without "this little instrument." The two men then discussed the negotiations about to take place in Genoa between Germany and the Soviets, and Rathenau was glad to learn that Kessler would attend the conference, albeit in an unofficial capacity.[17]

In the evening, Kessler was a dinner guest at the home of the Einsteins, recently returned from their trip to America. In his diary, he painted the following evocative portrait of his hosts, shortly before they departed for Japan: "Dined at the Einsteins' in the evening. A handsome, quiet flat in the west of Berlin (Haberlandstrasse 5); the meal, a little too plentiful and overly sumptuous, revealed a certain naïveté of this truly endearing, almost childlike couple." The other dinner guests were mostly members of Berlin's social elite,[18] and according to Kessler, the goodness and simplicity that radiated from his two hosts raised even these typical Berlin socialites above their usual level—the couple seemed almost patriarchal, and he was enchanted by them: "Einstein and his wife, whom I had not seen since their big journey abroad, answered my questions about the reception they received in America and England without embarrassment; their visit had indeed been a great triumph, with Einstein always giving a somewhat ironic and skeptical twist to the matter, saying that he really did not know why people took such interest in his theories. His wife said that her husband always told her that he felt like a swindler, like a con artist who does not give to people at all what they expected from him."

Einstein was to travel to Paris the following week to present a lecture at the Collège de France, and Kessler, who had recently returned from Paris, relayed to Einstein some detailed information from Paul Painlevé regarding relativity. Painlevé was, remarkably, both a mathematician working on the general relativity theory and the French minister of war.[19] A visit by a prominent German to France, almost four years after war's end, was still a very delicate undertaking, even though it was the avowed purpose of Einstein's visit to heal the rift between French and German scientists. Einstein had been reluctant to make the journey, and he had agreed to it only upon Rathenau's urgings. During dinner, Einstein told Kessler that his trip would probably raise suspicion in German academic circles as well, for these circles were "truly awful, and just thinking of them filled [him] with disgust." (Einstein assessed the reaction of German academics correctly, for when he returned from Paris, his colleagues at the academy gave him the cold shoulder.) He also let Kessler know that he had been invited to lecture in China and Japan and that he had told his wife, "As long as this ruckus persists, I have to see East Asia; that much, I must get out of it, at least."

As the other guests left, Einstein and Elsa kept Kessler back, and the three sat down on a sofa to chat. Kessler told Einstein that he sensed the significance of general relativity theory but did not really understand it. Einstein smiled and said that it was really quite simple, and he proceeded to give a lengthy explanation based on the shadows that bugs on the surface of a luminous glass sphere cast on a plane. When this failed to enlighten Kessler, Einstein pointed out that the theory's real significance resided in the connection it established between matter, space, and time: none of these three existed on its own, but each depended on the other two, and *that* was what was new in general relativity. But he failed to understand why people got so excited about it! When Copernicus deposed Earth from its central role in Creation, it produced an understandable stir, for it resulted in a radical change in how people perceived their place in the universe, Einstein said, but how had *his* theory changed anything in the way people perceive the world? His theory was indeed consistent with every rational worldview or philosophy. One could live with it whether one was a materialist, a pragmatist, or anything else.

*   *   *

Einstein's supposition that the ruckus attendant to his fame would soon fade away, and that he had best exploit it while it lasted, was, of course, quite erroneous. He remained an international celebrity for the rest of his life. His prominence did have a lot to do with the spirit of the times, because many people felt a desperate need to leave the disastrous war behind and to seek greater, less worldly visions. Furthermore, the emerging technologies of radio and film helped spread Einstein's fame rapidly and widely; they disseminated his words and his image all over the world. It did not hurt that Einstein was in addition a sympathetic and patient photographer's subject whose eyes seemed to exude wisdom, kindliness, humor, and a certain rebelliousness.

2.

Journey to the Far East (1922)

GETTING AWAY

On June 24, 1922, two young German Army officers, members of the ultranationalistic organization Consul, pulled alongside Walther Rathenau's open automobile as he was being driven to work and shot him to death. Rathenau's name, as well as Einstein's, had long been on the death list of monarchist and other extreme nationalistic associations, and his murder was but the latest of hundreds of political killings committed by them: a year earlier, Matthias Erzberger, another government minister, had been assassinated, and just one week after Rathenau's murder, the Jewish publicist Maximilian Harden, a severe critic of Wilhelm II, was attacked and seriously injured.[1] Einstein took heed and withdrew from his most controversial political activities.

A few months earlier, Kaizo-sha, a Japanese publishing firm specializing in progressive literature, had invited Einstein to present a series of public lectures in Japan. The idea for this venture had come from the firm's president, Sanehiko Yamamoto, and had a curious origin: Yamamoto had previously invited the renowned British philosopher Bertrand Russell to lecture in Japan and had asked Russell for the names of the three most significant persons alive, because he wanted to invite them to Japan also. Russell provided him with just two names, Lenin and Einstein, and since Lenin was otherwise occupied at the time, the invitation came to Einstein.[2]

The prospect of a long sea voyage and a tour of the Far East clearly tempted Einstein, not least because it offered an escape from the pervasive attention he was receiving in Berlin. The offer was, moreover, a handsome one: Einstein was to be accompanied by Elsa and would receive an honorarium of £2,000, less £700 for their travel expenses. With rampant inflation and a devalued currency holding sway in Germany, Einstein must have welcomed the opportunity to earn hard currency funds, which he needed for the support of Mileva and his two sons, who lived in Switzerland.

The chronicle of his tour in Japan shows that Einstein's substantial honorarium was very well earned. His contract with Kaizo-sha obliged him to adhere to a grueling schedule of travel, lectures, receptions, speeches, and dinners that often left him quite exhausted. One clause of the contract, furthermore, prevented Einstein from speaking at other public occasions. Yamamoto was, nevertheless, a most attentive host—as well as a very astute impresario and businessman, and he charged extremely high prices for tickets to Einstein's popular lectures.

Early in October 1922, Einstein and Elsa set out from Berlin for Marseille, where they would embark on their voyage. On their way, they stopped for several days in Zurich and Bern, where Einstein visited Michele Besso and Lucien Chavan, two old friends from his student days in Zurich.[3] The bitterness between him and Mileva over their breakup eight years earlier had diminished somewhat; thus it is likely that he also visited Mileva and his two sons, although he did not mention this in his diary. In the past year, Einstein had made several trips to Zurich to visit the boys, and he had taken them to Italy and on a sailing holiday on the Baltic coast. On those visits, he had stayed in Mileva's apartment and had invited old friends there for music-making—an arrangement that may not have been to Elsa's liking.[4]

On October 6, Einstein and Elsa left Zurich with all the luggage they would need for the next five months. Their overnight train to Marseille was overcrowded, but as they approached their destination, Einstein's main concern was how to move their luggage safely from the railway station to the harbor. Although the war had been over for four years, he knew from his recent trip to Paris that many Frenchmen were not ready to welcome German visitors. At the station, Einstein and Elsa were taken in tow by an honest-looking young man who nonetheless deposited them in a 'horrible tavern' where Einstein discovered a bug in his morning coffee. In a final indignity, they were transported to the harbor over exceedingly bumpy cobblestone streets, sitting on a luggage cart that lacked springs, before reaching the dock where their ship, the *Kitano Maru*, was tied up.[5] Once they were safely on board, Einstein gave the scoundrel (*Spitzbube*) who had been their guide an 'energetic' piece of his mind, presumably in French, and his tirade left the chap plainly offended.

SEA VOYAGE TO THE EAST

Einstein and Elsa were cordially welcomed by the officers of the *Kitano Maru*. Once he was comfortably ensconced in his cabin, Einstein's mood mellowed, and he set out to explore his new environment and meet his fellow passengers. He quickly made the acquaintance of Hayasi Miyake, a young, 'Europeanized' Japanese physician who was returning to Japan from Munich. Einstein would meet Miyake again during his last days in Japan. He also received a joyous greeting from a 'corpulent Russian Jewish woman' who welcomed Einstein as a fellow Jew. Almost all the passengers turned out to be either English or Japanese—a 'quiet, refined crowd,' remarked Einstein. When the ship cast off at noon on October 7, he was on deck and the sun was shining brightly. Since Einstein's new friend Miyake spoke German, the two chatted with each other as they watched Marseille and the hills surrounding the city recede into the distance.

At four in the afternoon, there was the obligatory lifeboat drill, and all passengers had to don their life jackets and assemble at their lifeboat stations. The drill gave Einstein his first opportunity to observe the Japanese crew; he found them to be friendly and precise, without being pedantic, but lacking in individuality. He concluded that the Japanese were unproblematic and impersonal and that they cheerfully fulfilled the social function assigned to them. They were unpretentious and took pride in their community and their nation, and even though they had abandoned their distinctive traditions in favor of European ways, their national pride was undiminished. He felt that the Japanese was, first and foremost, a social creature, possessing nothing so personal that it had to be hidden or locked up.

Einstein was keeping a travel journal for the first time, and he faithfully recorded each day's events—at least at the beginning of his long journey. Cut off from his normal personal and professional ties ashore, he turned to the universe that was observable from the *Kitano Maru* and kept a careful chronicle of the state of the ship ('licked clean' at four every morning) and of the weather, the sea, and the sky. His preoccupation with physics was not affected by the change in environment, however, and he worked on the unified field theory on a daily basis.[6]

Two days out of Marseille, the *Kitano Maru* passed the island of Stromboli, which presented a splendid sight bathed in morning sunlight. Einstein was particularly intrigued by the many small volcanic islands that jutted out of the sea near it, and he sketched one of them in his diary, with a cloud of water vapor hovering over its cone. Einstein was on deck as the ship passed through the Strait of Messina, and he was surprised by the 'severity' of the towns and of the landscape on both sides. The steadily rising temperature and the 'intoxicating air' put him in a reflective mood; he speculated that the Greeks and Jews of classical antiquity must have lived in a less debilitating climate, for it could not be coincidental that the zone of a flourishing intellectual life had shifted to the north since their time.

Einstein's Japanese fellow passengers fascinated him, particularly the mothers with their young children, and he liked to watch them scampering about on deck. Their astonished facial expressions reminded him of flowers.

Among the books Einstein had brought was Henri Bergson's latest philosophical work,[7] and it was the first book he read on the voyage. He credited the author with having a factual grasp of relativity theory, but he distanced himself from Bergson's philosophy. He turned next to a recently published book by Ernst Kretschmer, who is today remembered—if at all—for devising a system for relating mental illnesses to specific physical types. It is not known whether Einstein accepted Kretschmer's theory, but his book left a deep impression on him, on account of a depiction of a particular personality type: it jolted Einstein, for it mirrored his own. In a rare personal comment, Einstein summed it up this way: 'Hypersensitivity converted to indifference. In youth, introverted and unworldly. A sheet of glass between himself and others. Unfounded mistrustfulness. A paper ersatz-world. Ascetic impulses.' In a letter to Mileva that Einstein had written twenty years earlier, he had described himself in strikingly similar terms.[8]

Five days out of Marseille, the *Kitano Maru* arrived in Port Said, the northern terminus of the Suez Canal, where Einstein had his first taste of the exotic East. The harbor swarmed with 'boat loads of yelling and gesticulating Levantines of every shade,' who scrambled on board like creatures spewed out of hell. They made an earsplitting racket and converted the ship's upper deck into a bazaar, although no one seemed to be buying. The only ones who did well were some pretty, athletic boys who told the passengers' fortunes.

In Port Said, the *Kitano Maru* met a sister-ship. The patriotic fervor among the two crews that ensued persuaded Einstein that the Japanese were in love with their nation. The evening brought a spectacular sunset: the buildings and walls illuminated by the flaming red sky reminded Einstein of the intense colors often seen in paintings of tropical scenes. When he awoke the following morning, the ship was already in the Canal, traversing a vast expanse of brilliant yellow sand interspersed with small bushes, palm trees, and the occasional camel. After steaming through the green waters of Great Bitter Lake and the final section of the Canal, the ship reached the port of Suez, the southern terminus of the Canal. The town's pretty little houses, surrounded by palm trees, brought a welcome relief to Einstein's eyes after so much desert. Arab tradesmen arrived in sailboats and boarded the ship, but this time, Einstein perceived them as 'beautiful sons of the desert, robust, gleaming, black-eyed, and better mannered than those in Port Said.'

On October 14 the ship was underway again, heading south in the Red Sea. At sunset, a magnificent mountain chain came into view, silhouetted against the reddish-purple sky. It reminded Einstein, somewhat incongruously, of the Uetliberg, a small mountain and popular hiking destination just outside Zurich that had been visible from his office window at the ETH. In the star-bright night that followed, Einstein commented that he had never seen the Milky Way with such clarity and that it contained dark areas with sharp boundaries, some seemingly protruding from the galactic disc. (The Milky Way was at that time still the only galaxy known to astronomers.) The night was exceedingly hot, and Einstein kept comfortable by sleeping in the nude, with a fan.

After emerging from the Red Sea and the Gulf of Aden, the ship steamed east across the Arabian Sea. Einstein noticed that the sharks and flying fishes that had accompanied the ship in the Gulf had now disappeared, and, he explained to his diary, the greater depth (several thousand meters) of the sea did not permit sunlight to penetrate to the ocean bottom. As a result, there were little flora and fauna at the bottom of the sea, thus less life at the surface.

The ship was now approaching the equator, and the characteristic tropical weather patterns caught Einstein's attention. He noticed that with the rising temperature, the cloud cover increased. He reminded himself that at this time of year, the equatorial regions receive the most intense radiation from the sun, causing masses of moisture-laden air to rise and form clouds—and in his diary, Einstein sketched these weather phenomena. When these clouds release their water as tropical rain, air masses stream toward the equator from both north and south. These air masses are deflected by the rotation of the earth, giving rise to hurricanes and cyclones.[9] Einstein continued in his weather analysis: at other times of year, the regions that receive the most intense solar radiation shift northward or southward, giving rise to the whole complex of weather phenomena—the temperature rise being much greater over land than over the ocean. A few days later, off the coast of Sumatra, Einstein saw his first Fata Morgana mirage, and he resumed his meteorological musings: the ships seen near the horizon appeared to be floating in air, because the light that reached one's eye was refracted in an unusual way by the temperature and humidity gradients in the air.

One night, Einstein and the other passengers were rudely awakened by the ship's siren and thought there was an emergency. It turned out, however, that torrential rain had reduced the visibility to nearly zero, and the siren was sounded to warn other ships in the area.

Located between the ship's engine room and its sun-baked hull, Einstein's cabin was exceedingly hot. He suffered, moreover, from diarrhea and from 'horrible hemorrhoids' and had to seek relief from the ship's doctor. Neither his indisposition nor the extreme heat kept him from his work on a unified theory, but he reported no progress. When he did appear on deck, Einstein had to put up with being constantly photographed, either alone or surrounded by fellow passengers. It was a taste of what awaited him in Japan.

COLOMBO AND SINGAPORE

On October 28, the *Kitano Maru* approached her first port of call: Colombo, the capital of Ceylon (today Sri Lanka), then under British administration. The coastline was not yet in view when another heavy tropical thunderstorm forced the ship to drop anchor and wait for the squall to pass. Finally, at nine in the evening, a pilot arrived in a rowboat and came aboard to guide the ship into the harbor, where she docked next to another Japanese steamer. An elderly Sri Lankan—Einstein referred to the Sri Lankans as Indians—with a gray beard and fine, distinguished facial features came on board and implored Einstein for a tip upon delivering two telegrams to him. The "Indians" had skin color that ranged from brown to black, and while their faces and bodies were extremely expressive, their bearing was submissive. Because of this perceived combination of overwhelming pride and abject misery, Einstein compared them to noblemen turned beggars.

Einstein and Elsa had now been at sea for three weeks and looked forward to setting foot on land again. Early in the morning, they joined another couple in exploring Colombo's Hindu quarter and visited a Buddhist temple. They traveled in individual rickshaws drawn at a trot by 'Herculean, yet graceful individuals.' Einstein was deeply ashamed of being a guilty acces-sory to this revolting abuse of human beings. But there was nothing he could do about it, for swarms of these 'beggars with regal facades' pounced on every stranger until he capitulated to them. They were so skilled at begging and beseeching that they melted any heart.

Here, as later in China, Einstein was taken aback by the primitive living conditions of poor people. He blamed their deplorable situation on the trop-ical climate: it kept people from looking more than a quarter hour into the future or the past. Living on the bare earth, in the midst of filth and stench, they did little, and also required little; but under such crowded conditions, he felt that nobody had a chance of developing a separate individuality. On the other hand, there seemed to be no brutality, no screaming and yelling, such as he had witnessed in Port Said. These people were not market hawkers; they led a quiet, humble existence that was yet not lacking in certain merri-ment. Europeans suffered by comparison, mused Einstein, for they were softer, more brutal, and greedier. But, he went on, these were the very attrib-utes responsible for the practical superiority of Europeans and for their capacity to carry out great enterprises. Einstein was left wondering whether raised in this tropical climate, Europeans would not be like Indians.

The next morning, the harbor was alive with activity. Einstein watched the 'Herculean stevedores with their gleaming black bodies' unloading the ship's cargo. Elsewhere, youngsters performed diving stunts 'for the sake of filthy lucre' and for the amusement of passengers coarse enough to enjoy the show. At noon, the *Kitano Maru* left Colombo, and Einstein summed up his impressions of Ceylon: 'a floral paradise and a showplace of human misery.' He and Elsa would visit the island again on their return journey.

October 30 was the birthday of the emperor of Japan (the Mikado), and the big celebration that took place on the upper deck left the air ringing with numerous choruses of 'banzai' and the singing of the national anthem. The Japanese were evidently greatly moved by all this. Eerie chaps, thought Ein-stein, for whom the nation is at the same time their religion. The evening brought musical performances by the passengers; it was Einstein's first intro-duction to Japanese music. He compared one passenger's declamatory singing to the noise a tomcat makes when its tail is stepped on. Occasionally, the singer elicited a note from a guitar-like instrument, but with no discern-able connection to his singsong, which he accompanied with wild gestures. Having lived with Mozart and Bach all his life, Einstein found this music utterly incomprehensible at first. He specifically deplored the total lack of harmony or musical structure. A few days later, Einstein wrote of coming across a Japanese passenger singing for himself, an encounter that left him feeling nauseated.

One day, Einstein paid a visit to the captain on the bridge, a ritual he never omitted when at sea. He expressed an interest in the navigational instruments and was shown the ship's sextant, her chronometer, the speed gauge that was towed behind the ship, and her compass, which he considered to be rather primitive. It is perhaps not widely known that for many years, Einstein was deeply involved in the practical design of gyrocompasses, which were at that time—and for many years to come—the most sophisticated navigational instruments. Einstein had long been a friend and collaborator of Hermann Anschütz, whose eponymous company manufactured gyrocompasses.[10] Einstein made significant contributions to their design and held some patents.

Singapore was the *Kitano Maru*'s next port of call, and a curious task awaited Einstein there. The year before, he had accompanied Chaim Weizmann on a fundraising tour in America in support of the Hebrew University in Jerusalem, a project close to Weizmann's heart. As an avowed opponent of nationalism, Einstein did not, as mentioned, consider himself a Zionist, but he was happy to lend his support to the university. When Weizmann discovered that the ship Einstein was traveling on would made a one-day stop in Singapore, he saw it as an opportunity to obtain a donation for the Hebrew University: Einstein was to put the touch on Sir Menasseh Meyer, a fabulously wealthy Jew living in Singapore.[11]

After threading her way through the lovely green archipelago, the *Kitano Maru* docked in Singapore on November 2. Einstein and Elsa were welcomed by delegates of the city's Jewish community, who had been advised of Einstein's coming by Weizmann, and were then taken in tow by Alfred Montor and his wife, who drove them to their spacious home.[12] The Montors had evidently been selected to be their hosts ashore because they spoke German—his roots were in Hamburg; hers, in Vienna—but as Einstein soon discovered, they were not typical members of Singapore's Jewish community. The first Jews who settled in Singapore arrived in 1830 from Calcutta, the city many Baghdadi Jews had moved to during the harsh reign of Daud Pasha, ruler of Baghdad and Mesopotamia from 1817 to 1830. Some Sephardic Jews had lived in Baghdad since biblical times and spoke Arabic;

others had come following their expulsion from Spain, but Baghdad remained the spiritual home of Singapore's small Jewish community for a long time. With the advent of steamships and the opening of the Suez Canal in 1869, the importance of Singapore as a trading center grew, and European Ashkenazi Jews, such as the Montors, arrived to join the polyglot Jewish community. While the Ashkenazi and Sephardic Jews spoke different languages and did not socialize with each other, they did attend the same synagogue, where the services were held in Arabic. At the time of Einstein's visit, the Jewish community numbered a few hundred, of whom the majority were still Baghdadis.

Once Einstein and Elsa arrived at the Montors' home, they learned the full extent of how the 'indefatigable Weizmann' proposed to exploit Einstein's visit. Einstein's travel diary contains a detailed and somewhat bemused account of events during the next twenty-four hours. While speeches and hand-shaking sessions were hardly to Einstein's liking, he accepted them in good grace on this occasion, as he considered it to be for a good cause: extracting a sizable donation from Sir Menasseh Meyer. Einstein irreverently compared Meyer to ancient Lydia's fabulously wealthy king and referred to him as the Jewish Croesus.

On the afternoon of their arrival, Einstein and Elsa were driven to Meyer's palatial home on top of a hill that overlooked the city and the sea. Directly below stood a sumptuous synagogue, which had been especially built by Croesus, Einstein wrote, 'for communicating with Jehovah.' Einstein described Meyer as an erect, slender, strong-willed oldster of eighty, with a small, gray goatee; a narrow, reddish face; and an arched, narrow hook nose. He was clever, had rather sly eyes, and wore a small, black cap above his arched forehead. Einstein saw in Meyer a resemblance to the eminent Dutch physicist Hendrik Antoon Lorentz, whom he greatly admired, but unlike Lorentz's sparkling, benevolent eyes, Croesus's eyes were cautious and wily. His facial expression suggested systematic order and work, rather than—as in Lorentz's case—a love of humanity and a communal spirit. Einstein, ever responsive to female pulchritude, took greater delight in Meyer's daughter: she had a small, pale, patrician face and was 'the most beautiful Jewish woman' Einstein had ever seen.

Before the afternoon reception could begin, a long, drawn-out photography session took place, which Einstein, for once, seems to have relished. He was seated at the center, next to Meyer; the two men were surrounded by members of Meyer's family and numerous other married couples. Following this photo session, to which enormous importance was attached, he and Elsa entered a large 'oriental refreshment hall,' where a Malaysian band played Viennese waltzes and jazz in a 'European-schmaltzy coffee-house style.' Einstein sat between Meyer and the Anglican archbishop, a 'crafty, English-only speaking, slender, large-nosed English aristocrat,' who flirted—not without success—with Meyer's money, but without laying any claim on his soul. The conversation at the meal turned into a desperate 'linguistic calamity,' but Einstein found that the cake, on the other hand, was exceptionally tasty. The refreshments were followed by several ceremonial speeches and a reception at which Einstein was deeply moved by the genuine warmth of the Jews he met—despite enduring such an unremitting barrage of hand-shaking that he was reminded of his American journey.

On the way home, the Montors took Einstein and Elsa on a hurried tour of the city's bustling Chinese quarter, but there was not enough time to see, only to smell, remarked Einstein. In the evening, there was a gala open-air dinner at Meyer's home, at which some eighty guests were served a sumptuous meal and heard Einstein explain the needs of the new university in Jerusalem. The meal was hearty and endless, and, finally, Einstein had to leave the table because he couldn't bear to look at, let alone eat, any more dishes of delicacies. Once the meal was over, the Malaysian band reappeared and jazzed it up merrily, and everybody began to dance, including Meyer. At last, Einstein solicited Meyer in earnest for a contribution to the university. Despite subsequent attempts, Einstein could not discover whether 'one of [his] missiles had successfully penetrated Croesus' thick skin.' Apparently one did find its mark, and the Hebrew University in Jerusalem was enriched by five hundred British pounds.

That night, Einstein and Elsa slept under a mosquito net while it rained torrentially. After breakfast, they were taken on a tour of a rubber plantation before returning to the ship, where a group of 'devoted Jews' was waiting to bid them farewell.

After such a long time aboard with the same fellow travelers, Einstein was clearly pleased that the complement of passengers was 'enriched' by two laid-back Swiss officers and a youthful German businessman. During his brief stay ashore, Einstein had gained the impression that Singapore was almost entirely in the hands of the Chinese. He concluded that the Chinese were capable of supplanting every other nation, thanks to their diligence, their modest expectations, and their fecundity. On the other hand, he found it difficult to understand their mentality and he shrank from trying to do so, mindful of the extent to which Japanese music and singing had remained totally unintelligible to him.

HONG KONG AND SHANGHAI

During the weeklong passage from Singapore to Hong Kong, her next port, the *Kitano Maru* encountered a powerful typhoon and gigantic waves that 'made her dance.' It was Einstein's first potent storm at sea, and he watched the swarms of flying fish churned up by the ship. He also observed that having to balance oneself constantly was exceedingly tiring and that women were much more prone to seasickness than men.

The scenery he beheld upon entering the harbor of Hong Kong (on November 9) was the most beautiful he had seen on the entire journey: with the long, mountainous island alongside the rocky shore, the harbor between them, and the many small islands rising steeply from the sea, the scene reminded him of a 'half-drowned' Alpine landscape.

A delegation of Hong Kong's Jewish community had come to welcome him, but he excused himself from attending a reception they had planned. Instead, he and Elsa spent the day being driven around Victoria Island by two local Jewish businessmen. At midday, they stopped at an American-style luxury hotel for lunch. During the meal, their guides proved to be knowledgeable with regard to science and the Hong Kong colony, but they also let Einstein know what remarkable earthly delights the city had to offer.

The beauty and variety of the coastline, with its fjord-like inlets and floating fishing villages, delighted Einstein—but he was appalled by the living and working conditions of the Chinese he encountered. 'These tormented people, men and women, had to break stones or carry them for 5 cents a day, for this is how their heartless economic machine punishes the Chinese for their fecundity.' He mused that in their dreary condition, they may not even be aware of their misery, but that it was sad to observe all the same. On a more hopeful note, he added that he had been told that during a recent successful strike for higher wages, the Chinese laborers had been very well organized.[13]

In the afternoon, the Einsteins were delivered to the Jewish community center, which was set in a lush garden and offered a splendid panorama of the city below.[14] There they were joined by the wife and sister-in-law of one of their guides, and, over tea, Einstein learned that Hong Kong had only 120 Jews, and most of them were Arabic-speaking Baghdadis whose religious practices seemed to be even more fossilized than was the case with 'our Russian-European' Jews. Einstein nevertheless felt a real sense of kinship and belonging with these Jews from the lands of Euphrates-Tigris, because they were so similar to the Jews back home. Einstein inferred that Jews must have remained fairly 'pure' in the course of the last 1,500 [sic] years.

After tea, the party rode up to Hong Kong's so-called Peak, which offered a grandiose panorama of the harbor. The many islands rising steeply from the sea reminded Einstein of Alpine peaks shrouded in fog. The cable tramway they traveled on had been operating since 1888, with separate compartments for Europeans and Chinese, noted Einstein.

The next morning (November 11), Einstein and Elsa were taken on a tour of the Chinese quarter on the mainland, and the outing reinforced Einstein's impression of the Chinese: they were hardworking, dirty, and dispirited—even the children were listless. What a shame it would be, he mused, if all other 'races' were to succumb to the Chinese! The similarity in appearance of men and women surprised Einstein. He wondered, what was the fatal allure of Chinese women that left men defenseless against being blessed with so many children? In the evening, three Portuguese high school teachers came aboard to visit Einstein. In the course of their conversation, they claimed that it was impossible to teach the Chinese to think logically and, specifically, that they had no talent for mathematics!

Einstein left Hong Kong filled with admiration for the British administration, which he credited for the abundant, lush flora in the colony and for founding a university for those Chinese who opted for a Western way of life. The police force was staffed entirely by 'blacks' who hailed from India. The British were really admirable at governing, and they knew how to defuse the nationalist opposition by practicing tolerance—how different from the Continental Europeans, lamented Einstein.

On the morning of November 13, the *Kitano Maru* steamed up the Yangtze River estuary and arrived at Shanghai, where a party boarded to welcome the Einsteins: the German consul—a Mr. Pfister—and his wife, who were to be the Einsteins' hosts ashore; and the physicist M. Inagaki and his wife, who would be their very congenial guides and companions on the final leg of their journey to Japan. In Shanghai, it was the German community, not the Jewish community, that welcomed Einstein.

Following the obligatory press conference, with Japanese and American journalists asking the customary questions, Einstein and Elsa were taken to a Chinese restaurant. During the meal, a colorful and jovial funeral procession passed by—a Chinese custom that struck Einstein as almost barbaric. The meal, however, was highly sophisticated and endless. The constant fishing with chopsticks in communal dishes left Einstein's innards in a temperamental state all afternoon, but he managed to make it back to his haven, the Pfisters' home, in the nick of time.

Einstein and Elsa spent the afternoon on an extensive walking tour in the Chinese quarter, which left Einstein with much the same impression as in Hong Kong. The narrow streets were clogged with throngs of pedestrians and with rickshaws coated with every sort of filth. The air was filled with an infinite variety of stenches, and the streets were lined with open-air workshops of every sort. The Chinese crowds made a lot of noise, but Einstein never heard any quarreling among these 'gentle, dull and neglected people as they pursued their grueling struggle of existence.' Einstein's party also visited a theater where a different comedian performed on each floor, each leaving the audience grateful and delighted. On the street Einstein saw filth everywhere, but also many happy faces in the crowds. Even the coolies, who performed the work of horses, gave no sign of suffering. The presence of a group of Europeans occasioned a great deal of mutual gawking, and the Chinese gawkers were particularly intrigued by Elsa peering at them through her lorgnette—she was exceedingly shortsighted but, allegedly, too vain to wear glasses.

In the evening, the Inagakis drove Einstein and Elsa through a maze of dark streets to the home of the wealthy painter and calligrapher Wang Yiting, who had invited them to dinner. Behind the cold, high walls of his compound, they found a brightly lit, picturesque garden with a pond. In the festively illuminated halls, their host's own beautiful paintings were displayed alongside other, lovingly collected, works of art. The guests included the rector of Shanghai University and several teachers and other local notables, and the dinner was accompanied by countless syrupy speeches, one of them, admittedly, by Einstein. The excessively elaborate meal struck him as wickedly gluttonous and completely unimaginable for a European. Afterward, Einstein attended a reception at the Japanese Club, where the atmosphere was carefree, informal, and much more to his liking.

Next day, in the morning, Einstein and Elsa went on another sightseeing tour. This time, they visited a Chinese village and a Buddhist temple complex, where their presence again sparked a great deal of mutual gawking, before they hurried back to the ship. At three in the afternoon the *Kitano Maru* left Shanghai and headed east across the Yellow Sea to Japan. After steaming through the maze of small green islands in the Seto Inland Sea, she arrived in Kobe on the afternoon of November 17.

KOBE, KYOTO, AND TOKYO

In the months between the first announcement that Einstein was coming to Japan and his arrival, numerous books and magazine articles dealing with relativity theory were published in Japan, some in a popular vein, others more scholarly.[15] These had generated an enormous public interest in Einstein's visit, so that a large and enthusiastic crowd was waiting to greet him when he arrived in Kobe. His welcome there would, nevertheless, soon be dwarfed by what awaited him in Tokyo and elsewhere. The reception committee included the German consul and representatives of both the German Club and the Zionist organization; there was also a select group of Japanese physicists who were known to Einstein from their student days in Europe. They would accompany him on all his travels in Japan.[16] Following the mandatory press conference and the usual inane questions, Einstein and Elsa boarded the train for the two-hour trip to Kyoto, accompanied by 'the professors.' On their drive to the hotel in Kyoto, Einstein had his first close-up glimpses of Japan. He was enchanted by 'the magically illuminated streets with their neat little houses, the delightful schoolchildren, the simplicity and elegance of the Japanese—graceful, diminutive folk, tripping clap, clap, down the street.' The hotel was a massive wooden structure that overlooked the city. Once Einstein and Elsa were settled there, they dined together with the professors. The meal over, the professors still had ample enthusiasm for a scientific discussion session before they allowed Einstein's 'exceedingly strenuous' first day in Japan to come to a close.

Crowded into the following morning was a swift sightseeing tour of Kyoto. The Einsteins and the professors fitted in visits to some of the city's temples and gardens and to the famous ancient castle with its water-filled moat. It was then time to catch the train for the ten-hour-long journey to

Tokyo. They traveled in an observation car under a cloudless sky, and Einstein was captivated by the passing scenery—'the mountain passes, lakes and fjords, the lovely schools and the carefully cultivated land and exquisite clean villages.' He and Elsa were treated to a spectacular sunset just as the snow-covered Mount Fuji came into view. But scenic beauty was eclipsed when, soon afterward, a horde of journalists boarded the train, and the barrage of 'stupid questions' started all over again.

A huge crowd greeted the Einsteins at the station in Tokyo, where photographers let off so many magnesium flashes that Einstein was blinded for several minutes. After making their escape, he and Elsa were whisked to a reception at the Academy of Sciences, in which several German associations took part. Utterly exhausted, the pair were at last taken to their hotel, where masses of wreaths and bouquets awaited them. Two expatriate German academics, Sigfrid and Anna Berliner, came to call on them briefly, and then Einstein's day was finally done. He felt like a 'living corpse.'

Wilhelm Heinrich Solf, the German ambassador in Tokyo, submitted a report to the Foreign Office that describes Einstein's arrival and his lecture tour in more sanguine terms than Einstein did: "His journey through Japan resembled a triumphal procession . . . the entire population, from the highest dignitary to the rickshaw coolie, took part, spontaneously, without any preparation or fuss. At Einstein's arrival in Tokyo, such a huge throng was waiting at the station that the police were powerless to control the life-endangering crowding. . . . Thousands upon thousands of Japanese thronged to the lectures—at 3 yen per head—and [Einstein's] scholarly words were transmuted into yens that flowed into Mr. Yamamoto's pockets."[17] The ticket price cited by Solf, three yen, was ten times the cost of a meal.

Einstein had little time to recover from this tumultuous reception. On the very next day, November 19, he addressed an audience of two thousand people, in the first of his two public lectures. He discussed special and general relativity theory for four hours, and his words were translated, sentence by sentence, by the physicist Atsushi Ishiwara, who wore a 'picturesque Japanese costume' for the occasion. Einstein thought that it made him look like something between a penitent and a priest.[18]

The following day was only a little less hectic. Einstein was honored at a luncheon at the Academy of Science and gave a brief response. In the evening, he and Elsa dined in their hotel together with their host Yamamoto and several of his associates at the publishing house. Afterward they all went to a theatrical performance that included instrumental music, singing, and dancing. To Einstein's surprise, the members of the audience—mostly families with children—sat on the floor and took a lively part in the action. The stage sets were dazzling, and the acting was highly stylized, with men playing the female roles. The action on the stage was accompanied by a chorus of three men singing incessantly and by an orchestra that sat in a cagelike enclosure behind the proscenium. The music they played provided the rhythm and the mood of the action 'but was devoid of any structure or symphonic logic,' according to Einstein. Japanese music does indeed lack harmony, which has been the hallmark of Western music since the Middle Ages, and, evidently, harmony was of fundamental importance to Einstein's appreciation of music. After the performance, the party strolled through a brightly lit, but almost deserted, shopping street before repairing to a little restaurant for a chat. When Elsa and Einstein were finally back in the hotel, they found that the ever-thoughtful Yamamoto had arranged for fruits and cigars to be waiting for them in their rooms.

On November 21 the annual chrysanthemum festival took place in the gardens of the imperial palace. Einstein and Elsa were invited, but in getting ready for this important social occasion, a serious practical problem arose: how to locate the obligatory formal attire that would fit Einstein's large frame. With Yamamoto's help, a morning coat in his size was eventually located, but the top hat that came with it was much too small for Einstein's head, so that he was obliged to hold it in his hand all afternoon. They were picked up at their hotel by officials of the German embassy, who drove them to the imperial gardens. Einstein, along with the other invited dignitaries, lined up in a semicircle to await the appearance of the empress. She walked along the inside of the semicircle, exchanging a few words with some guests before stopping to chat for a while with Einstein, in French. All the guests then repaired to the meticulously landscaped gardens, where tables with refreshments had been set up and where many varieties of chrysanthemums were on display—all carefully arrayed, like soldiers, remarked Einstein. And, of course, there were countless hands eager to press Einstein's.

Ambassador Solf's official report of the same occasion is again more flamboyant than Einstein's: "The highest distinction was afforded the famous man at this year's Chrysanthemum Festival! It was not the Empress, nor the Prince Regent and the Imperial Princes who held court; everything turned subconsciously and involuntarily around Einstein. At this traditional celebration of the harmony between the Imperial family and its people, its 3,000 participants quite forgot the real significance of the day. . . . All eyes were focused on Einstein and everyone wanted to, at least, have pressed the hand of the most famous man of our time. An admiral in full uniform pushed through the throng, stepped in front of Einstein and said: 'I admire you' and went away again."[19]

Einstein, who could barely tolerate the formalities of German academia, who ignored all sartorial customs, and who clung to his own hair styling, had now been immersed in the rigid social rituals of a strange new culture for a whole week. He and Elsa must therefore have been greatly relieved when, following the imperial reception, they enjoyed a relaxed evening at the home of Sigfrid and Anna Berliner. Like many other German academics, these two had been unable to find suitable positions in Germany and had moved to Japan before the war. They now lived in a charming Japanese-style house in Tokyo. Although Sigfrid had studied physics and mathematics in Göttingen, he now held a professorship in economics at the Imperial University. Einstein was, in any case, more taken by his wife, Anna, who was a psychologist and, in his words, a gracious and intelligent woman—an authentic Berliner.[20]

The next day, Einstein was the focus of a festive reception given by Yamamoto's publishing firm. The reception took place on the street in front of the Kaizo-sha publishing house, and, naturally, it drew a large crowd of curious bystanders and assiduous photographers. Afterward the Einsteins visited a magnificent Buddhist temple decorated with striking wooden carvings, where priests presented Einstein with a precious art book. As Einstein and Elsa took their leave, they were again besieged by eager photographers in the courtyard, but, fortunately, the arrival of a group of merry schoolgirls from Osaka managed to cheer Einstein up.

Einstein's affection and admiration for his host Yamamoto had grown steadily ever since he set foot in Japan, and the next day, he and Elsa were guests for lunch at that 'splendid person's' charming home. How peaceful and undemanding these people must be, marveled Einstein, when he learned that Yamamoto's small house was home to three maids, one servant, and four students, aside from his wife and their young children! Together with Yamamoto, Einstein visited several Japanese homes and a farm in the afternoon, and he was fascinated by the simplicity and friendliness of the Japanese he met. He was particularly captivated by the young children, who were unfailingly cheerful and well-cared-for, and who were hardened against the cold.

Einstein next paid a visit to the president of the academy and discovered in the course of their conversation that the president's son had been a student of Professor Weber at the Polytechnikum in Zurich. Hearing the name of Weber, who had died ten years earlier, must have stirred sour memories in Einstein, for Weber had been his nemesis during his student days. Einstein had often skipped the ETH lectures and studied contemporary physics on his own, yet still obtained the highest final grade of the four graduating students. He was, nevertheless, the only graduate who had not been offered an assistantship. Bitterly disappointed, Einstein had blamed Weber's intrigues for his lack of success in obtaining a position at ETH or as a high school teacher. The experience had embittered Einstein and left him resentful for many years.[21]

On the same day, in the evening, Einstein was feted at a large reception and dinner at the German–East Asian Association. He was again surprised by the enthusiasm, friendliness, and sympathy for Germany that he encountered in Japan. He learned that the Japanese had established their own first-rate optical industry, thanks to the help they received from German engineers. During the reception, Einstein was drawn into so many different conversations with both German and Japanese guests that it set his head spinning like a carousel.

Because of Einstein's well-earned reputation as a defender of the powerless, strangers often solicited his support wherever he went. His reputation had apparently preceded him to Japan, for he spent the following morning studying the file of a certain Mrs. Schulze, the wife of an official at the German embassy, who was apparently being made a scapegoat to cover up a scandal. Einstein had been asked to intercede on her behalf. A few days later, a priest called on him to provide additional information about Mrs. Schulze's case, and after that, Mrs. Schulze's English physician came to assure Einstein that her psychosis was caused by the maltreatment she suffered at the hands of her husband. Einstein, who was by now on a friendly footing with Solf, wrote him a long letter concerning the affair, but it is unknown what the outcome was. Einstein was less troubled by a subsequent caller, whom he described as 'a crazy American lady who was convinced that she could cure other crazy people.'

Einstein met with a group of journalists on November 23, and for once, his conversation with them turned out to be an agreeable experience. Following a sumptuous meal, he visited a music school in the afternoon, where he attended a concert with the hope of gaining a better understanding of Japanese music. But after listening to 'melancholy flutes playing in unison, with many grace notes but no real melody' or harmony, he remained unable to come to terms with the music. The day ended with a relaxed dinner party and harmless conversations with Japanese physicists and a representative of the German embassy.

Einstein went for a long walk the next day, accompanied by his faithful companion, the physicist Inagaki, of whom he had grown very fond. The two had lunch in a tavern, where Einstein was impressed by the quiet good manners of the guests, but the roasted lobsters that were served there moved Einstein to pity for the 'poor creatures.' In the afternoon, he visited a private home that housed a splendid art collection, where he particularly admired ancient Japanese work. In Einstein's view, it mirrored the Japanese psyche better than the later, Chinese Buddhist–influenced art, which seemed remote and unrelated to the Japanese soul.

In the evening, at the physics department of the university, Einstein gave the first of a series of six daily lectures on relativity. It was a general lecture on the meaning of relativity and its consequences, four hours in length. Originally only some 120 faculty members had been invited to attend these technical lectures, but upon Einstein's urging, Professor Nagaoka permitted undergraduates to attend as well. Among the students was one who later reported that he had never seen such a lecturer: [Einstein] was always smiling and had great composure, and while he had a powerful intuition, he was not very good at calculations, nor rigorous in his logic.[22]

Einstein's remaining days in Tokyo followed a similarly busy schedule. After his second specialized lecture, he attended a musical performance that included a choir and some very young dancers. The entire corps of Japanese journalists then took him and Elsa to a guest house, where geishas danced to music and where a particular elder geisha's expressive, sensuous face left an unforgettable impression on Einstein. Eventually, he and Elsa were politely bade farewell so that the uninhibited second part of the night could begin. His curiosity aroused, Einstein later asked Inagaki to enlighten him regarding geishas and morality.

Between his daily lectures, Einstein found time to visit art museums, gardens, pantomimes, and concerts. He also attended a performance of a Noh play and was deeply affected by the masks and the slow pace of the performers' movements. But he was also obliged to take part in numerous receptions and 'devilishly sophisticated' dinners, which were always sprinkled liberally with speeches. On many of these occasions, Einstein was asked to play the violin, although he grumbled that his playing had deteriorated badly because he was so tired and had had no time to practice. At home in Berlin, he was used to playing almost daily.

Einstein was able to catch a glimpse of the contemplative Japanese lifestyle that he admired so much when he and Elsa were invited to a tea ceremony in an elegant Japanese home. Their host was the author of four fat volumes about the tea ceremony, and he showed them to Einstein with evident pride. Afterward Einstein was welcomed by ten thousand students at Waseda University, a university that continued to be run in the democratic spirit that its founder had introduced.[23] After the lecture, he met with a physicist who told him of the recently observed shifts of spectral lines from arc light sources.[24] After taking part in another welcoming reception by pedagogical societies, he was greeted warmly by a group of girls from a seminary who had been waiting outside: it was 'a charming, happy scene amid the jostling crowd in the dusk—too much love and indulgence for one mortal being.' Einstein was dead tired by the time he arrived back at the hotel. It had indeed been a long day, but not an atypical one.

The following morning, Einstein and Elsa were determined to strike out on their own, for once. They made their way to the railway station to obtain some travel information but encountered only a comical failure of communication. Heading back to the hotel, they were spotted by Inagaki, who had gone looking for them in a car, and he quickly had them in tow once again. The first event of the day was a three-hour-long concert of ancient traditional court music in the imperial chapel, the only institution where this archaic music was still being nurtured. The wonderful soundscapes produced by the plucked and wind instruments reminded Einstein of chorales. Byzantine and Chinese-Japanese music had common Indian roots, he was told. Following the concert, it was time for his afternoon lecture at the university, followed by a discussion with the local theoretical physicists that was, for once, substantive. Einstein was then celebrated by the delegations of twenty thousand Tokyo University students before he and Elsa were whisked off to a formal dinner at the German embassy. Although the affair was graced, as usual, by the presence of assorted diplomats and persons of eminence, Einstein enjoyed it because of the wonderful Western music he heard there: it offered a soothing reprieve after all those unfamiliar sounds. Einstein also dabbled a little on the violin—admittedly, very badly, because he was dead tired and had not practiced. But everything else at the reception was 'stiff and boring.'

In the last of his university lectures (on December 1), Einstein discussed what could be learned about the universe from general relativity theory— what he called the cosmological problem. At the end, he received warm thanks from the students, who were in the audience only because Einstein had interceded on their behalf. He was then given a gigantic farewell dinner in which Tokyo's entire intellectual elite took part. Einstein had to give a speech, and when he was politely compelled to play the violin, he performed Beethoven's Kreutzer Sonata.[25] It was Einstein's last day in Tokyo, except for an overnight stay later on, when he and Elsa were on their way south.

In the morning, he set out on the next leg of his Japanese odyssey, which took him to Sendai, a city some two hundred miles north of Tokyo. Elsa was left behind in Tokyo in the care of Mrs. Inagaki.

SENDAI, NIKKO, AND NAGOYA

Einstein's arrival in Sendai drew the usual life-threateningly large crowds to the railway station, and in a nearby hotel, he was welcomed at an official celebration. He gave a four-hour-long lecture the next morning, and in the afternoon, he was driven through the wonderful coastal landscape and visited an island that was covered with pine trees. On this trip, Einstein was accompanied by Yamamoto, Inagaki, and the caricaturist Ippei Okamoto, who would produce many excellent character sketches of Einstein in the days ahead.

On December 4, the four men were on the road again, traveling to Nikko by train. Einstein was awed by the mountainous landscape they passed through and took great pleasure in his three travel companions: 'splendid fellows, cheerful, modest, lovers of nature and art, unforgettable.' Nikko is about sixty miles from Tokyo and is famous for its complex of Shinto shrines and temples, many dating back to the eighth century. Among them is the mausoleum of Tokugawa, the father of the long and important dynasty that bears his name.[26] Unfortunately, the two ladies, Elsa and Mrs. Inagaki, who were to join the Einstein party in Nikko, had missed their train in Tokyo, but in the end they all met at the hotel.

Einstein, Inagaki, and Okamoto had planned to hike up to Lake Chuzenji (elevation 4,000 feet) the next morning, but Einstein discovered that Inagaki was loath to get out of bed, now that his wife had joined him. He and Okamoto therefore set out by themselves. The ascent took them through magnificent forests that presented splendid vistas, but soon their outing turned out to be far more challenging than they had bargained for: bitter cold and a heavy snowstorm caught them as they reached their destination, and these miserable conditions stayed with them all the way back down. Einstein's companion was worse off than he was, for the poor chap wore straw sandals; still, he remained cheerful and roguish the whole time.

In Nikko, for the first time since arriving in Japan, Einstein was free of the daily round of lectures and receptions. He relished the break. When telegrams from the German Society in Kobe pursued him to Nikko, he remarked that in Japan, at least, he preferred to deal with the Japanese: they were more like the Italians in temperament—relaxed and humorous—and were still completely steeped in their artistic tradition. On his trek to Lake Chuzenji with Okamoto, the two men had talked about the Buddhist religion and why so many educated Japanese were flirting with fundamental Christianity. Einstein was fascinated by how the Japanese had viewed the world before the nation was opened up to European influences not so very long ago. He was astounded to learn that there had been no awareness of why the sun's altitude was related to one's north–south location or why it was colder on the northern island (Hokkaido) than on the southern islands. It seemed to him that among the Japanese, intellectual urges were weaker than artistic ones, and he wondered whether that was due to a natural disposition.

The next morning, Einstein, Elsa, and their companions visited Nikko's temple compound, which is reached by ascending through cedar-lined paths and a system of courtyards. Einstein thought that the richly decorated buildings were somewhat overloaded, with the joy of depicting nature outweighing considerations of architecture—and even more, of religion. A priest delivered a long speech about historical matters.

Back at the hotel, Einstein and Elsa lingered after dinner to watch the setting sun illuminate the mountains, but then they had to pack in a frantic rush—and not without a row. They caught the train to Tokyo in the morning and spent the night in their former hotel. Since this was the last time they would be in Tokyo, they had to pack all their belongings and be ready to catch the morning train to Nagoya, their next destination. In the midst of this even more frantic rush, the couple quarreled once again. We know this because just like Samuel Pepys recorded marital squabbles in his famous diary, so did Einstein.

On December 7, after bidding a final farewell to their Tokyo hosts, Einstein and Elsa boarded the train, accompanied by Ishiwara, Inagaki, and Yamamoto and his wife. At Yamamoto's urgent request, Einstein spent the train ride writing down his impressions of Japan up to that point. What he wrote shows him as a remarkably sensitive observer of the Japanese psyche and of Japanese culture.[27]

The usual tumultuous crowd of students awaited the pair at the railway station in Nagoya, but with the welcoming festivities behind them, Einstein and Elsa enjoyed a relaxed evening meal in a tavern. In their hotel, Einstein ran into Leonor Michaelis, a biochemist he knew well from Berlin who was on lecture tour in Japan, like other German scientists during those lean postwar years. Michaelis was a pianist, and, needless to say, Einstein wasted no time arranging a music session with him for the following day.[28]

Nagoya is well known for its splendid castle, where Einstein took particular pleasure in the paintings of nature and court scenes that cover the castle's walls and doors. That afternoon, he played violin and piano sonatas with Michaelis, and in the evening, he delivered a popular lecture on relativity. He spoke before a large audience while standing in the center ring of an arena that was normally devoted to sumo wrestling (*kokugi-kan*). The hall was unheated, and Einstein had no choice but to wear his overcoat throughout the lecture.

Einstein enjoyed pipe smoking, and when his supply of pipe tobacco ran out on December 8, he set out on his own the next morning in a futile attempt to replenish it. He walked along Nagoya's main street as far as the railway station before Yamamoto, Ishiwara, and Inagaki found him. They brought him to a renowned Shinto shrine situated in a large grove that contained many temples and elegant wooden buildings; while Einstein was impressed by the architecture, he also sensed that the pervasive idolatry, ancestor worship, and emperor cult were being exploited by the state.

KYOTO AND OSAKA

Einstein and Elsa were now again in Kyoto, a beautiful town with innumerable temples and gardens. After the usual warm welcome by the local physicists and students, Einstein proceeded to lecture for four hours on the philosophy of relativity theory, speaking in a splendid but exceedingly frigid hall, with Ishiwara translating his words sentence by sentence. After the lecture, he visited Kyoto's famed imperial gardens and the coronation castle, where he found that the courtyard's architecture was particularly impressive. He quipped that in the coronation hall, the gods seemed to be putting on imperial airs. What surprised and moved him most deeply was the portrayal of forty Chinese statesmen who were revered by the Japanese for bringing Chinese cultural fertilization to Japan. He remarked that reverence such as this seemed to be very much alive still, as witnessed in the touching veneration that Japanese students accord their teachers. He had even been told that some-where in Japan, there was a temple dedicated to the bacteriologist Robert Koch.[29] Einstein felt that such serious reverence without a trace of cynicism or skepticism was typical of the Japanese; that they were pure souls to be found nowhere else. 'One cannot help loving and revering this country.'

On December 11 Einstein made a day-trip to Osaka, a large manufac-turing and commercial city. The mayor and many students greeted him at the station and escorted him to a huge reception at a hotel, where he was intro-duced to the local dignitaries and had to press countless hands. The gigantic communal meal that followed was accompanied by military trumpet music and by many speeches, all of them dripping with pathos, and one, admittedly, by Einstein. Then, in the evening, Einstein lectured before an audience of 2,500 people, and though it had been a full day, he was not terribly tired because everyone had been so considerate and modest. Upon his return to the hotel in Kyoto, he found Elsa seriously outraged for having been left at home.

Einstein had no obligations the next day and was able to catch his breath. He visited the ancient Nijō Castle of the Tokugawa shoguns in the afternoon. But then his mind turned to physics, and he spent the remainder of the day cal-culating the energy tensor of an electromagnetic field inside an isotropic mass.

The next morning, he was off on another one-day trip, this one to Kobe, and this time, he wisely invited Elsa along. When they arrived, they joined Yamamoto and a rising young politician for lunch in a fishing village outside Kobe. It was then time for Einstein's usual four-hour lecture with Ishiwara, which did not end until eight in the evening. Einstein and Elsa left the lec-ture hall quickly to attend a dinner at the German consulate, which was fol-lowed by a reception in Kobe's German Club. By then it had become so late that they had to return to Kyoto on the milk train and did not get back to their hotel until one in the morning.

Kyoto University gave a festive luncheon for Einstein the next day (December 14), with all professors in attendance. At the student assembly that followed the luncheon, a student delegate extended a genuinely warm greeting to Einstein, speaking in perfect German. Einstein had been asked to devote his Kyoto lecture to discussion of the genesis of relativity, and he obliged by presenting the only authoritative account of how he arrived at the relativity theory. Fortunately, Ishiwara took notes during the lecture, and his notes were eventually translated and published.[30] That evening, Professor Akihiro Nagaoka arrived from Tokyo, bringing with him a suitcase full of splendid gifts for Einstein.

Einstein and Elsa remained in Kyoto for the next four days, and with no lectures or receptions scheduled, they had a sorely needed opportunity to recover from their strenuous itinerary. Yamamoto's choice of Kyoto as their recuperation site was a wise one: the town had been Japan's capital from 794 until 1869, and it was full of magnificent shrines and temples—a tourist's paradise. In Kyoto, Einstein was also able to catch up on his voluminous correspondence and to continue working with Ishiwara on their joint paper. One day, he accompanied Elsa to a silk shop, where they saw wonderful weavings, but apart from that, he spent most of the time visiting the city's temples and museums. At a particularly splendid Buddhist temple (Chion-in), where the priests gave him a warm welcome, he was intrigued by the temple's enormous bell, which employed a horizontal suspended log as an exterior clapper. At twilight, he and Elsa were taken to another celebrated Shinto temple, Kiyomizu-dera, which is perched on a steep mountainside, supported by tall wooden poles. Afterward, they all strolled down a festively lit shopping street—a happy scene, remarked Einstein, with many people carrying paper lanterns and waving little flags. It reminded him of the Oktoberfest in Munich.

Together with his constant companion Inagaki, Einstein climbed several hills to view the sunset and to witness the striking light effects in the maple forests. The two men also made an overnight trip to nearby Nara, where they visited the temple compound and where Einstein had the curious experience of being sniffed by a herd of tame deer. He admired the architecture of Nara's Great Eastern temple (Todai-ji), said to be the world's largest wooden structure,[31] and the giant statue of the Buddha it housed, but he was increasingly bothered by the overabundance of superstition he encountered everywhere, including all those slips of paper (*omikuji*) fluttering from trees and shrines.

FAREWELL TO JAPAN:
MIYAJIMA, FUKUOKA, AND MOJI

When Einstein and Elsa boarded the train to Miyajima on December 19, it was the start of the last leg of their Japanese tour. Miyajima is a small sacred island off the coast, a few miles below Hiroshima. It is best known for the Itsukushima shrine, which is supported on poles in the floodplain, and the floating gate (*torii*), which is surrounded by water at high tide. After an all-night train ride, the Einsteins arrived on Miyajima in the early morning. They flopped into bed and slept until ten, then went for a walk along the bewitchingly beautiful coast until noon. Later in the afternoon, Einstein and Inagaki scaled Mount Misen, the mountain that dominates the island and affords a spectacular vista of Japan's inland sea. The trail leading up to the 1,500-foot-high summit has steps hewn into the granite and leads past countless shrines dedicated to a variety of nature gods—testimony to the Japanese's love of nature and their dedication to every sort of harmless superstition, in Einstein's view.

Amid these idyllic surroundings, Einstein was jolted back to reality and drawn into the political turmoil back home. He received a cable from Ambassador Solf informing him that Maximilian Harden, the German-Jewish journalist who had narrowly escaped assassination soon after Rathenau's murder, had stated in court that Einstein had gone to Japan because he no longer felt safe in Germany. In his telegram, Solf asked Einstein for permission to deny that allegation. Einstein wired back that this issue was too complicated for a telegram and that a letter was called for instead. That evening he composed his response, in which he advised Solf that Harden's comment was unwelcome to him because it complicated his situation in Germany and that it was, moreover, not quite right, but neither was it quite wrong. In the opinion of people with a clear perspective of the prevailing situation in Germany, his life had indeed been in danger in the aftermath of Rathenau's murder. The murder had changed his view of the situation, but what had actually prompted him to accept the invitation to visit Japan was his desire to see the country and, partly, a need to escape the tense atmosphere at home that had so often placed him in difficult situations. Solf forwarded Einstein's letter to the Foreign Office in Berlin, along with a lengthy account of the adulation with which Einstein was welcomed in Japan, adding how much he, personally, had come to admire Einstein for remaining down-to-earth, gracious, and modest in spite of it. At the end of his report, Solf remarked that Einstein was now on his way home, and he requested that all possible assistance be extended to Einstein upon his return to German territory. Einstein's luggage would contain various honorific presents, wrote Solf, and his wife had requested that these objects be admitted duty-free.[32]

During the next few days on Miyajima, Einstein took long walks along the shore, accompanied by Inagaki and Okamoto. In their hotel, the entire party survived an episode of carbon monoxide poisoning caused by the open charcoal fire, an incident that Einstein passed off as minor but which had seriously affected Elsa and Mrs. Inagaki. Einstein and Elsa next traveled to the town of Moji, on the southernmost tip of Honshu island, where their arrival was celebrated with a lavish reception. After the obligatory press conference, they were installed in princely fashion in the exclusive Mitsui Moji Club.[33]

The following morning (December 24), after submitting to being 'photographed for the 10,000th time,' Einstein traveled by train and ferry to Fukuoka, the major city on the southern island of Kyushu. He arrived barely in time for his four-hour-long lecture before an audience of over three thousand people. In the evening, he was feted at a seemingly endless dinner, at which, to Einstein's manifest surprise, most of the dignitaries became thoroughly intoxicated and gleefully animated.

Hayasi Miyake, the physician Einstein had met on board the *Kitano Maru* at the start of his voyage to Japan, was his host in Fukuoka; it was Miyake's hometown. He took Einstein to a Japanese-style hotel for the night, and when they arrived, the hostess fell on her knees and welcomed Einstein by touching her head to the floor 'about a hundred times.' Einstein stayed in a suite of exquisite rooms, but what most impressed him were the sliding paper doors that could comfortably be opened or closed using just one's little finger. One of his rooms had thoughtfully been furnished with European seating.

Einstein regarded the next day to have been a wild one, although it really seems hardly different from many others. Inagaki and several other academics arrived in his suite at nine in the morning and were soon followed by the 'droll' hotel hostess. She brought with her six large squares of silk, a bottle of India ink, and paintbrushes, and she had Einstein paint his name on the silks. When he was done, he met with Ayao Kuwaki, the first person to introduce Japanese physicists to relativity theory, and discussed epistemological implications of relativity with him. Einstein and Inagaki then rushed to the railway station to meet their respective wives arriving on a train from

Moji. The two couples explored Fukuoka and its shops together before heading for a formal luncheon at the medical faculty, which was accompanied by speeches by the president and Einstein. Following the meal, Einstein was showered with presents, shook the hands of a large number of professors, and was then shown exhibits that had been specially set up for him. These included gallstones, histological specimens with spirochetes, and crossbred fishes.

Einstein then visited the home of Hayasi Miyake and met his four enchanting children.[34] Later he was taken to an exhibit of paintings, specially mounted for his benefit by the provincial governor, before it was time for the return trip to Moji. Naturally, *everybody* came to the station to bid him and Elsa farewell, including that 'most genial of all Japanese hostesses.' By that time, Einstein commented, he was dead and only his corpse returned to Moji. No sooner had he arrived, however, than he was dragged off to a children's Christmas party, where he was obliged to play the violin for the children. When he finally returned to the Moji Club at 10:00 p.m., he ate, answered many letters from back home, and went to bed.

Einstein spent his remaining three days in Japan more quietly. He climbed yet another hill with Inagaki to enjoy the splendid panorama of sea and mountains. Later that evening, a man appeared with a sheaf of papers and asked Einstein, to his astonishment, to write down for him his impressions while on the summit. Then Yamamoto arrived and, deeply embarrassed, informed Einstein and Elsa that they would have to move, because the Mitsui Moji Club was 'showing its fangs' and had made horrendous monetary demands.

The next day, Einstein and Elsa went on a memorable all-day sightseeing cruise on a Mitsui steamer in the Inland Sea. Afterward they went for a leisurely walk through the town of Shimonoseki before they were feted at an exceedingly animated farewell dinner with Yamamoto and several others. They all insisted that Einstein paint his name on pieces of silk, and he was glad to oblige. Everyone was in very high spirits, and Einstein felt relaxed and at ease with his Japanese hosts, to whom he had become very close during the preceding five weeks—'blissful peace!' (*eitel Friede!*) During the dinner, Einstein presented Mrs. Yamamoto with a poem and a drawing—such was Einstein's customary thank-you. He composed similar doggerel verses on many other occasions all his life, and they were invariably deftly rhymed, witty, and perceptive.

On his last full day in Japan, Einstein was the guest of Moji's Commercial Club, the 'haut finance,' as he put it. It was his first opportunity to get together with Japanese persons who were not connected with the academic establishment. The businessmen he met on this occasion struck him as shrewd, not as polished as the professors, and more like their European counterparts. He appreciated their simple, straightforward manner. Throughout the meal, the businessmen gave individual singing performances, and Einstein was, naturally, obliged to play the violin.

3.

Homeward Bound:
Palestine and Spain (1923)

SHANGHAI, HONG KONG, SINGAPORE, AND COLOMBO, REVISITED

On December 29, 1922, Einstein and Elsa boarded the *Haruna Maru* in Moji and began their long journey home. Seeing them off at a poignant farewell party were Yamamoto and Ishiwara with their respective wives; Kuwaki with his young sons; Ishiwara; Miyake; and several gentlemen from the Mitsui-NYK shipping line, owners of the *Haruna Maru*. Once aboard, Einstein found the ship to be large and comfortable—she was certainly younger than the *Kitano Maru*, which had brought him and Elsa to Japan.[1] The ship left port at four in the afternoon and steamed westward across the Yellow Sea. Einstein soon settled into his comfortable cabin and turned to his work. He made progress in some electrodynamics calculations and wasted no time writing a letter to Ishiwara to let him know. Before leaving Moji, Einstein had received several back issues of Frankfurt newspapers, sent to him by his friends Sigfrid and Anna Berliner. Now he had an opportunity to read them, and after catching up on the depressing news from Germany, he lamented to his diary, 'woeful Europe!'

The weather was beautiful when the Einsteins arrived in Shanghai on the last day of 1922. They were met at the dock by two gentlemen, Mr. De Jong and Mr. Gatou, whom Einstein referred to as 'the engineer and the *parvenu*.' They learned that Gatou, the *parvenu*, would be their host in Shanghai, and although Einstein regarded him as a snob, he gave him a lot of credit for owning an excellent piano, something that Einstein had sorely missed in Japan. In the evening, Gatou hosted a noisy New Year's Eve celebration in his home. Despite finding himself seated next to a charming Viennese woman, in a rare personal comment Einstein confessed in his diary that he felt sad.

Shanghai depressed Einstein. The Europeans he and Elsa encountered during their two days ashore seemed 'lazy, self-confident and shallow' to Einstein, and they all employed numerous Chinese servants. On New Year's Day, in the afternoon, Gatou gave a reception in honor of his famous house-guest, and the endless speeches and handshaking dismayed Einstein, who characterized the guests as 'syrupy philistines.' In the evening, he and Elsa were taken to a popular Chinese place of amusement, where he discovered to his amazement that the Chinese played European music indiscriminately, for any occasion—a wedding or a funeral—regardless of whether the piece was a funeral march or a waltz, as long as it had plenty of trumpets. According to the resident Europeans, the Chinese were 'dirty, tormented, dull-witted, good-natured, dependable, gentle, and surprisingly, healthy.' United in their praise of the Chinese, they also asserted that the Chinese lacked business sense, offering as evidence the fact that Europeans were able to compete successfully with them, even though Europeans' salaries were ten times higher!

The *Haruna Maru* left Shanghai on January 2. Einstein was delighted to recapture the peace and quiet of shipboard life. He was so anxious to main-tain that happy state that he avoided making any new acquaintances. It seems unlikely that Elsa did likewise, but since Einstein's diaries rarely mention her activities, little is known of them. After enduring so many weeks of crowded schedules, it is not hard to see why Einstein savored what he called his 'con-templative, enviable existence' on board. At work in his cabin, he continued to search for ways to generalize Eddington's recent theory.[2]

When their ship arrived in Hong Kong (on January 5), Einstein and Elsa hoped to have a little peace and privacy at last and snuck ashore secretly, early in the morning. For once, they managed to elude the welcoming recep-tion. They planned to take care of several errands ashore, the first being a visit to the office of the Mitsui-NYK shipping line—where they promptly ran into Mr. Gobin, one of the businessmen who had driven them around Victoria Island on their first visit. They had been treated so hospitably on that occasion that they could not refuse Gobin when he insisted that they come to an impromptu reception at the Jewish community center that he would arrange for the afternoon.

Gobin accompanied Einstein and Elsa all the way to the French consulate, their next destination, where they applied for their all-important transit visas. They then took leave of Gobin as quickly as they could. On their own again, they took the cable car to Victoria Peak, as on their previous visit. From there, Einstein climbed all the way to the summit in spite of the great heat. He was rewarded by a magnificent vista of the harbor, the sea, and the many islands. Einstein and Elsa, evidently enjoying the use of their legs again, decided to descend on foot from the Peak by following a trail that took them through a lush tropical grove. The descent took about an hour, and along the way they encountered a constant stream of Chinese men, women, and children, who groaned as they lugged bricks up the steep trail. The Chinese were the most unfortunate people on Earth, Einstein ruminated: they were cruelly abused and treated worse than cattle—their reward for being so modest, gentle, and undemanding.

The Einsteins had barely returned to the ship when 'that fellow' (Gobin) arrived to drive them to the Jewish community center, next to a synagogue. A comical scene awaited them there: almost nobody had shown up for the reception! As a result, they felt obliged to accept Gobin's invitation to dine with his family, and after surviving an 'endless, dreadfully seasoned meal,' the couple finally managed to get back to the ship.

When the *Haruna Maru* left Hong Kong harbor the next day and headed south, Einstein was sitting on the sunny deck. Wearing a hat, he found the heat to be just bearable, as he looked out on the scores of sailing junks dancing on the waves. The scene gave him a fresh idea regarding the problem of electromagnetism in general relativity—but he did not reveal in his diary what it was. Most likely, as had often happened, it turned out to be flawed. The weather grew increasingly overcast, warmer, and more humid, as the ship steamed south through the South China Sea, and after an uneventful passage, she dropped anchor off Singapore on the evening of January 10.

Just before he left Japan, Einstein had received word that he had been awarded the 1921 Nobel Prize in physics, but he had not thought that this news warranted mention in his diary. His disdain for prizes and honors is well known, but the prize money was another matter, for it had been assigned to Mileva in her divorce contract with Einstein. After his return, Einstein assisted her in investing the funds wisely. The prize was awarded for his work on the photoelectric effect, because the Royal Swedish Academy of Sciences deemed the theory of relativity to be still too controversial.[3] The news had not come as a surprise to Einstein, for von Laue and Svante Arrhenius, a member of the Swedish Academy, had both dropped broad hints before he left Berlin that his presence in Stockholm, in December, would be most desirable. Nevertheless, he had declined to change his travel plans.

Upon Einstein's arrival in Singapore, he received congratulatory messages from Arrhenius, Niels Bohr, and Planck, and he responded to them. His exchange with Bohr is particularly appealing because it illustrates the warm relationship between these two giants of twentieth-century physics, in spite of their philosophical differences regarding the interpretation of quantum mechanics.

Bohr had experienced his own annus mirabilis in 1913, eight years after Einstein's, when he proposed a quantum theoretical model for the hydrogen atom. Originally, Planck had postulated that the energy of the "radiative oscillators" was quantized, in order to explain the black-body radiation spectrum; Einstein had then introduced light quanta to explain the photoelectric effect and later showed that in a solid, the vibrational energy of atoms was quantized, as well. At this stage (1913) Bohr showed that the structure of atoms and their spectra also had to be understood quantum theoretically. These three men, Planck, Einstein, and Bohr, are rightfully considered the fathers of quantum physics.

In his letter, Bohr congratulated Einstein on the Nobel Prize; he recognized that the honor meant little to him, but thought that the funds attached to the prize might ease his working conditions. (Bohr judged Einstein's indifference to the honor correctly but was unaware that the prize money would go to Mileva.) Knowing that he too was being considered for a Nobel Prize (he received it the following year), he wrote Einstein that he was glad he would not be honored before Einstein's contributions to his work were recognized.

Responding in a lighthearted vein, Einstein wrote that he was touched by the genuinely Bohr-esque fear that he might get the Nobel Prize *before* Einstein; he wrote that Bohr's latest work had accompanied him on this voyage and had made him love Bohr's mind even more. He had loved Japan and found a sea voyage like this to be fabulous for a "ponderer" like himself—as good as being in a monastery. Finally, he informed Bohr that he was writing from near the equator and that tepid water was dripping languidly from the sky, spreading serenity and drowsiness over everything—as his letter bore witness. He signed it: A. Einstein, who reveres you.[4]

On the morning that the *Haruna Maru* docked in Singapore harbor,

Alfred Montor, the Einsteins' host on their earlier visit, was waiting for them. He drove them to a preserved primeval forest, where Einstein was deeply impressed by the wild and impenetrable profusion of plant life. They lunched at the Montors' home, and then visited a palm tree plantation before returning to the ship. Later that evening, Einstein summed up his day ashore very succinctly: 'Trees magnificent, human beings banal.'

Einstein called on Sir Menasseh Meyer ('Croesus') the next day. Unfortunately, there is no record of what passed between the two men, so eminent in their own spheres and so different from each other. The visit did, however, provide Einstein with another opportunity to admire Meyer's beautiful, noble daughter, whom he referred to as 'Portia' in his diary.[5] Having thus fulfilled their social obligations, Einstein and Elsa enjoyed the spectacular vistas offered by Singapore Island on the drive back to the ship. At five in the afternoon, in the midst of a tropical downpour, the *Haruna Maru* left the dock and threaded her way through the archipelago of 'bright green, velvety islands.'

The *Haruna Maru* now steamed north along the western coast of Malaysia, making two brief stops along her way. First, she dropped anchor off Malacca in the early morning of January 13, but the Einsteins were unable to begin their sightseeing outing until afternoon, when they were rowed ashore. They visited a Portuguese church and strolled around the town, fascinated by its lively mix of Indian, Malay, and Chinese inhabitants. The Malaccans traveled around town in two-wheeled carts equipped with straw roofs and drawn by long-horned water buffalos. Although it was tremendously hot, Einstein went back to work after returning to the ship— and discovered a devastating error in his latest theory. As he put it: 'I found a thick hair in my electricity soup. Too bad' (*Schade*).

The next day, they reached Penang, some three hundred miles farther up the Malaysian coast. The *Haruna Maru* dropped anchor in the bay, quite far from the town. It was again brutally hot on board, and as soon as they could, Einstein and Elsa went ashore, not only to sightsee but to escape the heat, which they found more bearable in town than on board. Although they were pursued by rickshaw men, they persisted in wandering around Penang on foot. The town's buildings, its boats, and its people had real style, in Einstein's view. He was particularly taken by a strikingly beautiful and very insistent beggar woman. In Penang, he and Elsa also visited a Buddhist temple, which was decorated with glaringly colorful paintings that Einstein found mysteriously terrifying; also a mosque with slender white minarets and an attached public bath, where several men lounged. Toward evening they returned to the ship, along with several Japanese fellow passengers. The sea had become quite choppy, and when their boat was tossed about, Elsa became very frightened—but not enough to keep her from scolding the sinewy Indian oarsman with his flaming black eyes. Standing upright in the stern of the boat, he calmly rowed on and delivered everyone safely to the ship. The heat was unrelenting till midnight.

Underway again, the *Haruna Maru* headed west across the Bay of Bengal to Ceylon. Einstein was on deck enjoying the pleasant breeze under a brilliant starry sky and chatting with a Singhalese school teacher about life in Ceylon. The teacher had nothing but high praise for the English administration of the island. Back in his cabin, Einstein continued his search for a unified theory. Although his work had suffered many reverses, he extolled the enviable existence he enjoyed on board.

When the *Haruna Maru* arrived in Colombo (on January 19), Einstein tried to organize an automobile party among the passengers, undoubtedly in the hope of avoiding another demeaning rickshaw ride. When that plan came to naught, he and Elsa went ashore by themselves, and after riding around town in a tram, they walked to the railway station, continually pursued by a host of importuning natives. They took a train to Negombo, a small coastal town some thirty miles north of Colombo, where, they had been told, there were no Europeans. Upon arriving in Negombo, they gave in at last and hired two rickshaw men. One of them was a totally naked 'natural man,'[6] while the other had once worked as an elephant handler in the famous Hagenbeck circus in Hamburg, a city that he could not praise highly enough.[7] The men drove Einstein and Elsa along Negombo's main street, which was lined with individual small houses amid a palm grove. Naturally, the pair of strangers was gaped at a great deal, but no more than a Singhalese couple would be in Berlin, commented Einstein.

The rickshaw men took Einstein and Elsa to a fishing village where the children were completely naked and the men wore only loincloths. Einstein's interest was piqued by the primitive outrigger fishing boats he saw there. He guessed that they were very fast, but also quite uncomfortable to sit in. A boat was returning, loaded with fish, and trailed by a flock of envious ravens. He would later sketch such a fishing craft in his diary. He and Elsa then passed a bay where a small stream entered the sea, and there, in a meadow beside the stream, they saw an enormous crocodile only about thirty feet away. The beast had already been spotted and was the target of a barrage of stones thrown by yelling villagers, as it slowly ambled off toward the water.

Einstein and Elsa dined at a guest house in the village before being brought back to the railway station. The Hamburg rickshaw man was so delighted with Einstein and with the five rupees that Einstein gave him that he returned to the station and presented Einstein and Elsa with a bunch of bananas for their train journey back to Colombo. While waiting for the train on the platform, they made the acquaintance of a gorgeous young Singhalese woman, accompanied by her sister and mother: 'village aristocrats,' opined Einstein. The young woman's great-grandfather had been Dutch, and Einstein declared that he had rarely seen anyone as beautiful as she was.

Inside the railway carriage, Einstein and Elsa were plagued by swarms of mosquitoes that caused them some concern, for the train was passing through a rice-growing region that was known to be infested with malaria. As soon as they arrived at the station in Colombo, they were again set upon by rickshaw men, and after resisting resolutely for a long time, they had to capitulate for the drive back to the dock. Einstein concluded that in these parts, it was considered impertinent (*eine Unverschämtheit*) for a European to go anywhere on foot. When they were rowed back to the ship, the sea was again so choppy, and Elsa so terrified, that she heaped bitter reproaches on her husband.

On January 22, the *Haruna Maru* put out to sea and headed west across the vast Indian Ocean. Her passengers would not see land again until they approached the Arabian Peninsula. Einstein never seemed more content; he relished the serenity on board and the brilliant starry night skies. He completed the manuscript on Eddington's ideas on general relativity that he would mail to Planck from Port Said, where he and Elsa planned to disembark to begin their expedition to Palestine. On their last day at sea, they joined the captain at a Japanese farewell dinner. There was also a masked ball at which several of the passengers gave musical performances, something the Japanese did with great virtuosity. Despite Einstein's reputation as a loner, and contrary to his resolve to act the misanthrope on board, he had become friendly with a number of his fellow passengers, among them several Japanese businessmen, a Greek ambassador returning from Japan, and a sympathetic English widow who insisted, over Einstein's protests, on presenting him with a British pound, for the University of Jerusalem.

Einstein was deeply troubled by the news that French troops had occupied the Ruhr region, the industrial heartland of Germany, to punish Germany for falling behind in her war reparation payments. He deplored the fact that in the past hundred years, the French 'had not become cleverer.'[8]

Once the *Haruna Maru* reached the Red Sea, the sunsets were particularly spectacular. As the sky changed from yellow-orange to deep red, some of the small jagged islands they encountered were brilliantly illuminated while others were left darkly silhouetted, a panorama so striking that it affected Einstein profoundly. When the ship approached Port Suez, the sea took on a deep blue hue and seemed remarkably transparent. Einstein wrote, 'The sky slightly overcast. Silvery muted colors. Picturesque sailing craft. Yellow shore.'

After a brief stop in Port Suez, the *Haruna Maru* entered the Suez Canal. Einstein remained on deck and was deeply stirred by the delicately lit, desolate scenery that unfolded as the ship made her way north and traversed the Great Bitter Lake with its 'barren deserted shoreline, wonderful.'

PALESTINE

They reached Port Said, the northern terminus of the Canal, early in the morning on February 1. Einstein and Elsa went ashore, along with their considerable baggage, and quickly passed through the customs formalities thanks to the kind assistance of their new friend, the Greek ambassador. They were well rested and looking forward to the new and very personal adventure lying ahead: a two-week tour of Palestine.

As much as Einstein had been delighted with Japan and all things Japanese (Japanese music excepted!), it was his visit to Palestine that affected him most deeply. To appreciate the state of affairs he encountered there, it is useful to recall the historical developments that preceded his visit.

Palestine had been occupied by many of the great powers of the Middle East since antiquity. The most recent was the Ottoman Empire that had wrested the territory from the Mamlūk sultans in 1517. Exactly 400 years later, the Ottoman Empire collapsed following the defeat of its forces in Palestine, in World War I. The famous Balfour Declaration, issued as a letter from the foreign secretary Lord Balfour to Lord Rothschild, declared that His Majesty's Government viewed "with favor the establishment of a national home for the Jewish people in Palestine, it being clearly understood that nothing was to prejudice the civil and religious rights of existing non-Jewish communities in Palestine."

The Balfour Declaration was endorsed by the Allied powers and the United States in 1922, and the League of Nations entrusted Great Britain with the mandate to govern the region. Later that year, the portion of Palestine east of the Jordan River was split off and became the Hashemite Kingdom of Jordan, while the remaining territory was administered by a British high commissioner. Sir Herbert Samuel, Einstein's host in Jerusalem, was the first occupant of that office.[9]

Jewish settlers from Europe had arrived in Palestine in appreciable numbers beginning in 1882, many of them having escaped pogroms in Russia and Poland. Although these immigrants encountered many hardships, including heavy Turkish taxation and the hostility of their new Arab neighbors, there were about ninety thousand Jews living among some six hundred thousand Arabs in Palestine at the time of Einstein's visit. Most of the Jews lived in communal agricultural settlements (kibbutzim and moshavim).

The Einsteins' journey from Port Said to Jerusalem turned out to be complicated and strenuous. Guided by a young man sent to meet them at the customs house, Einstein and Elsa boarded a train that took them south along the western bank of the canal to Candara (El Qantara), where a ferry carried them across. It was eleven at night before they finally boarded the train that was to take them to Palestine. After numerous delays, the train steamed eastward, following the Mediterranean shoreline. Einstein was surprised and delighted when the helpful young train conductor turned out to be a Berliner who, before he emigrated to Palestine, had attended several rallies in Berlin at which Einstein had spoken.

The train traveled through a desert landscape before crossing into Palestine in the early morning. Einstein sat at a window as the train rolled through a plain with very sparse vegetation. Eventually, a few Arab villages and Jewish settlements came into view, each surrounded by a grove of olive or orange trees or cacti. When the train reached the railway junction of Lod, several of the early Zionist leaders, known to Einstein from Berlin, climbed aboard to accompany him and Elsa on the last leg of their journey.[10] From Lod the train chugged along to reveal the spectacular rocky valley that leads up to Jerusalem.

Einstein and Elsa had been invited to be the guests of Sir Samuel, the high commissioner. On arriving in Jerusalem, they were met by an army officer who drove them to Samuel's official residence—Samuel's Castle, as Einstein called it. Einstein thought it an enormously pretentious stone building, but situated on the summit of the Mount of Olives, it offered unparalleled views in all directions. The building had originally been conceived by Kaiser Wilhelm II during his 1899 visit to Jerusalem, and it was intended to serve as a hospice for German pilgrims in Jerusalem. Its chapel contained a mosaic mural depicting the kaiser, with his wife by his side, holding a replica of the building. 'Authentically Wilhelmesque,' observed the bemused Einstein.[11]

While this was the first meeting between Samuel and Einstein, the two men soon became friends and would remain so for many years, during which they kept up a lively correspondence.[12] Einstein described Samuel as a many-sided, well-educated, and erudite Englishman, whose high-mindedness was moderated by his sense of humor. Sir Samuel's household included, apart from his wife, their unassuming adult son Edwin; his earthy and high-spirited wife, Hadassah; and their little son.

Despite the overcast sky on his arrival, Einstein was delighted by the panorama that Samuel's Castle offered of the city, of the hills surrounding it, and of the Dead Sea and the mountains beyond. His delight only increased the next day (February 3), as he and Samuel walked (it being the Sabbath) along a footpath past the city wall and past white stone houses, many surmounted by cupolas; the sun was shining brightly in a clear blue sky—an enchantingly beautiful sight. Joined by the secular philosopher Asher Ginsberg, the trio entered the old city of Jerusalem through an ancient gate and strolled through the narrow streets and bazaars. They visited the raised temple square where Solomon's temple had once stood. A magnificent sight, thought Einstein, who particularly admired the Dome of the Rock, reminiscent of a Byzantine church, but considered the basilica-like Al-Aqsa mosque to be in mediocre taste. The three men visited the Wailing Wall, where orthodox Jews faced the wall and prayed aloud, their bodies swaying back and forth. A deplorable scene, thought Einstein, of 'dull-witted fellow-members of the tribe,' men with a past but no present. He and Samuel continued their stroll through the old city, 'teeming with diverse holy men and races, very filthy, noisy and with a strange oriental air,' before taking a magnificent walk along the top of the city wall.

Einstein had lunch with several resident German Jews, including Asher Ginsberg, Arthur Ruppin, and Hugo Bergmann.[13] Their table conversation ranged over both happy and serious topics, though Einstein did not record the details. They were delayed by a horrendous rainstorm, after which Einstein visited a gloomy old synagogue in the Jewish quarter, where a group of 'scruffy religious Jews' awaited the end of the Sabbath, immersed in prayer. The Sabbath over, Samuel and Einstein were driven back on a road that had been turned into a sea of mud by the heavy downpour—apparently, an almost daily occurrence.

The following day, Ginsberg and Hadassah, Samuel's very capable and 'merry' daughter-in-law, took Einstein on an automobile trip. They drove through the soft, bare hills down to the Jordan valley and to the ancient ruins of Jericho. They stopped at a lush tropical oasis, and after lunch in the 'Hotel Jericho' they continued over incredibly muddy roads along the broad Jordan valley to the Jordan Bridge. Here a group of 'magnificent Bedouins' made an impression on Einstein. He and Elsa spent a relaxed evening at home, chatting with Samuel and Hadassah. The 'unforgettable splendor' Einstein had witnessed that day moved him profoundly. He sensed the matchless magic of this severe, monumental landscape and of the 'dark, elegant sons of Arabia, in their rags,' who inhabited it.

In the morning, Einstein visited a Jewish settlement that devoted itself not to agriculture, but to the construction trades. Most of the newly arrived settlers lacked any prior experience in the building trade, but after just a short period of training they performed superbly. The foremen who supervised the work were elected by the workmen and received the same salary as the others. In the afternoon, Einstein met first with the philosopher Hugo Bergmann, who had taken on the formidable task of establishing the Hebrew University's library; he met later with a local mathematics teacher, who showed him some remarkably interesting results in matrix algebra.

Undoubtedly, the high point of Einstein's day occurred that evening, when he had a chance to play chamber music, something he had greatly missed while in the Far East. His hosts were Norman and Helen Bentwich. At the end of the evening, Einstein asked their forgiveness for letting the music go on far too long; he pleaded extreme hunger for music (*weil so ausgehungert nach Musik*).

It is fortunate that an independent account of that evening exists. It was written by the hostess, Helen Bentwich, and is contained in one of her weekly letters to her mother in England. Her husband, Norman, was an army officer stationed in Palestine, serving as attorney general in the British administration. The Bentwiches lived in Jerusalem from 1919 to 1931; the following excerpt is from one of Helen's letters.

[11 February 1923] The great event has been Einstein. Monday evening we went to an 'At Home' at Ussishkin's to meet him. He is very simple & bored by the people but very interested in the music provided for him. Mrs. Einstein is a mixture between a Hausfrau and a Madonna. Tuesday evening they came to dine, & there was music. Margery, Thelma, Norman, a man Feingold & Einstein played a Mozart quintet. Norman on the viola & Einstein on Norman's violin.[15] He looked very happy whilst he was playing, & played extraordinarily well. He told some interesting things about Japan and his visit there, and talked of music, but not of his theories. He said of some man—'he is not worth reading, he writes just like a professor'— which was rather nice. He only talks French & German, but his wife talks English. She said they got so tired of continual receptions & lectures, & longed to see the interesting places they visit alone & simply. Wednesday afternoon he gave a lecture. About 250 people were there, including government officials, Dominican Fathers, missionaries, & of course a lot of Jews. He gave it in French, which of course was a handicap to me to start with. But, although I understood more or less every consecutive argument, the fitting them together was the trouble.[14]

Local politics undoubtedly played an important role in Einstein's decision to deliver his lecture not in German, the language most familiar to his audience, but in French. The lecture took place in a hall of the British police academy on Mount Scopus and was widely heralded as the first scientific lecture delivered at the nascent Hebrew University. Einstein was persuaded to begin his lecture with a greeting translated into Hebrew, which he laboriously recited. After the lecture, Samuel thanked him with some witty comments to the audience before taking Einstein away on a leisurely walk and engaging him in philosophical conversation.

The next day's sightseeing tour took Einstein to the Church of the Holy Sepulcher and the Via Dolorosa, and later he was given a rousing welcome by Jerusalem's Jewish community. He suffered through a reception attended by all of the city's notables and conversed with them on 'learned and other matters.' In his diary, he observed that by the end of the evening he was feeling 'completely at ease with all these charades.'

Einstein was driven to Tel Aviv on February 8 and spent several hours visiting Palestine's first high school (*Gymnasium*) before being taken to yet another reception at City Hall to receive the honorary citizenship of the city. During the tour of the city that followed, he inspected its central power plant and an agricultural research station and he was honored by Tel Aviv's engineering society. After dinner, Einstein attended yet another gathering with the 'Learned Ones' (*mit den Gebildeten*), at which he gave a speech. Einstein was astonished by what had been accomplished in this young, modern Jewish city and by its bustling economic and cultural life: 'Incredibly vigorous folk, our Jews!'[16]

Einstein's Zionist hosts were well aware of his ambivalent attitude toward Zionism, and they arranged visits to a wide range of different settlements and individuals during the next several days. They evidently hoped that if he gained a favorable impression on that tour, he might become more amenable to living and working in Palestine.

The first settlement Einstein visited was Rishon Le-Zion, which was also the first modern Zionist settlement in Palestine. Founded in 1882 by ten Russian pioneers (*Chaluzim*), it survived and eventually prospered thanks largely to the substantial financial support it received from Baron Edmond de Rothschild. Rishon Le-Zion is today well known for the excellent wines produced by its vineyards. Einstein visited its agricultural school and wine cellars and commented that the settlers led a joyful and healthy life, but were economically not yet self-sufficient.

Einstein's next stop was Haifa. Accompanied by Asher Ginsberg and the physicist Aharon Czerniawski, he traveled by train past scores of Arab and Jewish settlements to Jaffa, where workers of the local salt works came to the station to greet him. Then on to Haifa, but by the time the train arrived, the Sabbath had begun and there were no taxis at the station. As a result, Einstein had to slog his way over tremendously mucky roads to the home of the Pevzners, his hosts for the night. Mr. Pevzner was the brother-in-law of Hermann Struck, a renowned German Jewish artist and friend of Einstein from Berlin—from where Struck had recently immigrated to Palestine.[17] Einstein enjoyed a tranquil dinner at the Pevzner home, and it did not escape his notice that his host's wife was a 'dainty and keenly intelligent' woman. Afterward, a crowd of bland visitors arrived 'out of curiosity'; still, Einstein found the evening bearable, thanks to the presence of Struck and his wife.

The Strucks were, in fact, very congenial, and Einstein spent the next day in their company. They took a long walk on Mount Carmel, admiring the view of the port and city below. Their next stop was a high school, where they met its effective, albeit 'Prussianized' principal, Arthur Biram, another early German Zionist. Einstein had lunch in the Strucks' apartment, taking pleasure in the relaxed table conversation. After paying a brief visit to Chaim Weizmann's mother, who was surrounded by numerous 'sons, daughters, etc.,' Einstein went for another walk with Struck to visit an Arab author who was married to a German woman.

Einstein was unusually at ease in Palestine, in part because he had fewer formalities to deal with, and also because he was able to converse in German with most of the people he encountered. However, he could not escape academic ceremonies entirely: in the evening, he was honored at the still nascent Technikum (now, Technion). There were many speeches—one by Einstein, and two outstanding talks by Aharon Czerniawski and Elias Auerbach.[18] The ceremony ended with communal singing of psalms and of Eastern European songs by candlelight.

The next morning (February 11), Einstein visited first the workshops of the Technikum—primarily an engineering school at the time—and later a new and ingenious, almost fully automated oil refinery. Together with his two guides, Einstein then headed for Lake Tiberias (Sea of Galilee), stopping along the way at Nahalal, a collective settlement (moshav) being built according to the plans of the German architect Richard Kaufmann.[19] Its settlers, almost all Russian Jews, owned individual parcels of land, but the settlement's administration and the construction of buildings were communal. It rained, and it grew dark as they drove through the mountains to Nazareth and from there to the guest house at the Migdal farm, where Einstein would spend the following three nights.

The afternoon's heavy downpour quickly turned the road into a river of mud. When they finally arrived at the guest house, the sun had long set and their car was being drawn by a pair of mules. Once there, Einstein was compensated for the strenuous journey with an opulent meal. He spent a relaxed evening with his German host, whose family had returned to Germany, having had their fill of the hard life here. Einstein cheerfully put up with the primitive conditions on the farm. On his 'droll pilgrimages' to the outhouse, he carried a large lantern to light his way.

In the morning, the sun was shining brightly. Einstein was driven to Lake Tiberias, where he was astonished by the lush landscape along the shore, complete with boulevards lined with pines and palm trees—a scene that

Einstein as a young—and very self-possessed—boy, ca 1884. *Ann Roman Picture Gallery/ HIP/ Art Resource.*

Einstein and his sister Maja, ca. 1885. *Ann Roman Picture Gallery/ HIP/ Art Resource.*

Einstein and Maja, ca. 1893. *Courtesy of the Leo Baeck Institute.*

Einstein (front row, left) with his graduating class in Aargau, 1896. *Bildarchiv Preussischer Kulturbesitz (BPK)/ Art Resource.*

Einstein in the patent office, Bern, 1905. *Ann Roman Picture Gallery/ HIP/ Art Resource.*

Einstein and Mileva in Bern, ca. 1905. *adoc-photos/ Art Resource.*

Einstein walking with his stepdaughter Margo and her husband Dimitri Marianoff. *Bundesarchiv.*

Elsa Einstein at a peace demonstration in Berlin, 1919. *Bundesarchiv.*

Einstein on board the Rotterdam, 1921.
Library of Congress.

Five physicists, Nobel laureates all, in Berlin, 1928: Walther Nernst, Einstein, Max Planck, Robert A. Millikan, and Max von Laue. *Bildarchiv Preussischer Kulturbesitz (BPK)/ Art Resource.*

Menachem Ussishkin, Chaim and Vera Weizmann, Einstein, Elsa, and Ben Zion Mossensohn, at the start of Einstein's first visit to America, 1921. Ussishkin and Mossensohn were Zionist officials. *Library of Congress.*

Einstein and Elsa on board the Rotterdam, 1921. *Library of Congress.*

Elsa and Einstein meeting with President Harding in the garden of the White House during Einstein's first trip to America, 1921. *Library of Congress.*

Walther Rathenau, 1921. *Bundesarchiv.*

Einstein's long-term friend Hermann Anschütz-Kaempfe, 1930s. *Bildarchiv Preussischer Kulturbesitz (BPK)/ Art Resource.*

Einstein and Hans Ludendorff at the Potsdam Observatory, during an astronomy conference, 1929. *Bildarchiv Preussischer Kulturbesitz (BPK)/ Art Resource.*

reminded him of Lake Geneva. The area, however, was malaria-infested, and thus farming presented a continual challenge to the Jewish settlers. Back at the guest house, he met an exceptionally beautiful young Jewish woman at lunch, as well as an interesting, erudite laborer. Einstein's next destination was the picturesque town of Magdala, where, he was told, Mary had been born (actually, it was Mary Magdalene) and where Arabs were selling land to archeologists at 'horrendous prices.' His next stop was the communal settlement of Deganyah, Palestine's first kibbutz. Located near the point where the Jordan River emerges from Lake Tiberias, most of its settlers were again Russian. Einstein was impressed by their zeal, toughness, and idealism, which they retained in spite of being continuously beset by hunger, disease, and debt. Einstein did not believe that the communism they practiced would last forever, but he did think it would succeed in fashioning complete human beings. On their way home, the weather was fair and he and his guides took the opportunity to pay a quick visit to the picturesque town of Nazareth— before the heavy rains began again. Einstein was happy to get back to his cozy German guest house.

February 13 was Einstein's last full day in Palestine, and he returned to Elsa and 'Samuel's Castle' in Jerusalem, passing once more through Nazareth and Nablus. The weather was hot when he left the guest house, but it turned uncomfortably cold along the way. Torrential rains began once again. A truck had become stuck in the resulting morass, forcing vehicles and passengers to make their strenuous way, separately, through fields and across ditches. Automobiles are sorely taxed in this land, reflected Einstein.

That same evening, Einstein gave another lecture. The auditorium was packed, and this time he spoke in German. Afterward came more of those inescapable speeches, followed by a diploma from the association of Jewish physicians. In presenting the award, the speaker was stricken with stage fright and fell utterly silent. Praise the Lord, mused Einstein, that there are some among us Jews who are not so self-assured. At the reception that followed, his hosts closed ranks and pressed him with great urgency to move to Jerusalem—but it was in vain. Einstein was firm: 'The heart says yes, but reason says no.'[20]

ON THE ROAD AGAIN

Einstein and Elsa left Samuel's Castle for the railway station in the early morning. They were accompanied by Hadassah, who traveled with them as far as Lod, where they boarded the train to the Suez Canal. Elsa, who had developed a high fever the night before, felt worse and worse, and by the time they reached Candara, she collapsed completely. Fortunately, a kindly Arab train conductor assisted her and Einstein. Several officials in Candera provided Elsa with eggs and with a place to rest until ten at night, when she and Einstein crossed the Canal and boarded the train to Port Said. When they arrived, the two weary travelers were fortunate to find shelter and a warm welcome in the home of Mr. and Mrs. Moushli. The long and stressful journey to Port Said was a dismal ending to their Palestinian expedition, but Einstein expressed his confidence that Elsa would recover and that everything would turn out all right.

By the next day, Elsa had indeed improved. She had been nursed with great devotion by Frau Moushli and was clearly recuperating. Einstein, meanwhile, went for a walk along the jetty at the Canal's entrance, to the gigantic statue of de Lesseps, the Canal's architect, standing at its end.[21] The sun was shining brightly, and the row of bathing huts lining a nearby beach reminded Einstein of a colorful cubistic painting. He felt a sense of liberation (*Gefühl von Befreiung*).

On the morning of February 16, Einstein and Elsa, now fully recovered, boarded the steamship *Oranje* for the final leg of their journey, to Spain. The *Oranje* was a much older and smaller ship than the *Haruna Maru*, and her passengers were almost all 'English colonials.'[22] Einstein pronounced the food on board to be abominable, and in combination with heavy seas in the Mediterranean, it made him quite ill. When the *Oranje* docked in Naples, Vesuvius was hidden by gray clouds. The weather was so cold and unfriendly that Einstein was happy to remain on board.

The news from Germany continued to be bad. Einstein was deeply concerned about the possible consequences of the French railway strike and of the repressive French occupation of the Ruhr region; he sighed to his diary: 'where will it end?' When the *Oranje* docked in Toulon, the people were friendly enough, but not so in Marseille, where, Einstein was told, it was dangerous to be heard speaking German. The freight manager at the Marseille railway station refused to ship Einstein's luggage to Berlin and would not even ship it to Zurich.

Finding himself back in troubled Europe and coping with his complicated itinerary must have taken a toll on Einstein. For almost the first time,
he failed to make daily entries in his diary; instead he lumped together the
events of several days and reported them in telegraphic style. Eventually, the
entries petered out entirely.

SPAIN[23]

Einstein and Elsa had been glad to leave the *Oranje* in Toulon. They traveled
by train to Marseille and to Barcelona, where they arrived on February 22.
They were to stay in the city for a week, where a very crowded—and
stressful—schedule awaited them, so much so that Einstein made hardly any
entries in his diary. He nevertheless enjoyed his stay in Catalonia tremendously, remarking, 'it was lovely!' (*Schön wars!*)

Einstein's invitation had been arranged by the mathematician Esteban
Terradas and the engineers Serrate Lana and Rafael Campalans, all of whom
served as his academic hosts in Barcelona.[24] All three had been students in
Switzerland and Germany and spoke German fluently. Campalans, who was
also an active Socialist politician and a champion of Catalan nationalism,
took Einstein to a performance of Catalan national dances and folk songs.
Afterward the two men went to the Refektorium, a restaurant popular with
Catalan nationalists, where Einstein consumed a café au lait. Einstein, whose
political views were well known here, was also courted by Spanish anar-
chists, who saw him as a hero. They welcomed him at a meeting of the Sindi-
cato Único, an organization that had been linked to political violence. His
attendance at an anarchist gathering raised eyebrows, even though he had
declined to address the group. His views were widely misquoted in the press,
but the ensuing political discussion remained muted.

Einstein delivered three technical lectures in French at the Institut d'Estudis
Catalans. Although his lectures were nominally restricted to persons with a scientific background, they were, in reality, attended by an overflowing audience
of whom few understood what he said—apart, that is, from the white-bearded,
bald-headed, bespectacled mathematicians who sat in the two front rows.
According to this newspaper account, Einstein spoke slowly and observed the
faces of the listeners before him, their brows knitted and wrinkled by the torture
of incomprehension and the difficulty of following Einstein on his discon-

certing flights.[25] At another lecture, at the Academy of Sciences, Einstein gave a less specialized talk in which he discussed the philosophical and cosmological implications of relativity. He was also honored at a reception at City Hall and at other diverse ceremonial affairs. Amidst all this, he found time for two sightseeing trips outside the city: to the twelfth-century Cistercian monastery at Poblet and to Terrassa's stunning Romanesque Cathedral Basilica.

In Barcelona, Einstein and Elsa were often in the company of the local German consul and his wife, Ilse von Tirpitz. She was the daughter of Grand Admiral Tirpitz, the man who had been the chief architect of Wilhelm's Imperial Navy.[26] She and her husband, and the Terradas couple, were in the small group that gave Einstein and Elsa a warm send-off at the railway station when they boarded the night train to Madrid.

When they arrived the next morning (on March 1), they were met by the physicist Blas Cabrera, an old acquaintance of Einstein, who was his host in Madrid.[27] The welcoming committee also included the German ambassador, members of Madrid's German community, and delegations from the Faculty of Sciences and the College of Physicians. Also waiting for them at the station were Einstein's cousins Lina and Julio Kocherthaler, who made their home in Madrid; they drove Einstein and Elsa to the Palace Hotel, their residence for the next ten days.[28] At the hotel, Einstein huddled with the local notables over the busy schedule that had been prepared for him. Cabrera then took Einstein to see his laboratory and told him that he planned to take him and Elsa to a concert in the evening; but Einstein opted for something typically Spanish, and so they went to a musical revue instead.

The next day, Einstein gave the first of his three lectures at the university, in which he discussed special relativity. Though attendance was again nominally reserved for persons with a science background, the audience consisted of the most important politicians as well as scientists and many others who came to see the great man. Few in the audience comprehended the words of the lecturing Einstein, who strolled around as he spoke, one hand in his pocket, occasionally halting, deep in thought, to gaze at the ceiling. Directly after the lecture, Einstein was taken to a banquet in his honor, at which the elite of Madrid's medical community was in attendance.

On Sunday morning the Kocherthalers drove the Einsteins around Madrid until it was time for Einstein to prepare his response to the speech that Cabrera would give at the Academy of Sciences later that day. The session was chaired by King Alfonso XIII, who presented Einstein with a diploma of membership of the academy, at a ceremony witnessed by the cream of Spanish society and intelligentsia. In his response, Einstein praised the introductory speech of the academy's president, Carricido (*wunderbare Rede*), and then launched into a public exchange with Cabrera: they reviewed the current state of physics, with Einstein stressing the need for a new mechanics that was capable of dealing with quanta. (Quantum mechanics was still two years away.)

Later in the afternoon, the Marquesa de Villavieja feted Einstein and Elsa in her palatial home, at a "tea of honor." The most distinguished members of Spanish nobility and of academia had been invited, to encourage mingling of the "aristocracies of blood and of intelligence," according to one newspaper report. It was also reported that Einstein and the violinist Antonio Fernández Bordas performed an improvised "intimate concert," and though Einstein professed little taste for the social life, he was observed to be smiling, while "seated on an ample sofa, surrounded by cushions and women."[29] In his diary, Einstein dismissed this glittering social occasion with the remark: 'Then tea with an aristocratic society lady.'

On Monday, at a special meeting of the Mathematical Society, Einstein responded to written queries about relativity that were submitted by its members. Einstein then paid a visit to Santiago Ramon y Cajal, whom he admired (*wunderbarer alter Kopf*), considered to be the first neuroscientist because of his detailed work on the brain. It was then time for Einstein's second lecture at the university. Speaking again in French, he occasionally had to ask for help from members of the audience to find the French equivalent of a German term—help that was readily provided by the many German-speakers.

By constructing a web of 'many lies,' Einstein and Elsa managed to evade their very solicitous hosts the following day. This freed them to join the Kocherthalers, the art historian Manuel Cossío, and the writer Ortega y Gasset on a one-day excursion to Toledo.[30] Afterward, Einstein confided to his diary that this excursion turned out to be 'one of the most beautiful days of my life.' The sun was shining brightly, and Toledo seemed to him like a fairytale city. 'An enthusiastic old gentleman' (Manuel Cossio) guided the group through Toledo's medieval streets and plazas: from the river Tajo with its stone bridges, to the Cathedral, to an old synagogue with a small garden that afforded a magnificent panorama. They also visited a small church, where an El Greco painting of a nobleman's funeral moved Einstein 'more deeply than anything he had seen.'[31] All in all, it had been 'a wonderful day.'

The next day, Einstein had an audience with King Alfonso XIII and the queen mother at the Royal Palace. The king's dignity and simplicity impressed Einstein, but his mother's attempts to display her scientific knowledge suggested to Einstein that no one had ever told her what he really thought. That afternoon, in the final lecture of his visit, he spoke of the latest developments in general relativity theory. Einstein felt certain that almost no one in the audience, who listened so reverently, understood what he said. In the evening, he was honored at a reception given by Ernst Langwerth, the German ambassador, and while Einstein took a liking to Langwerth's family—they were lovely, uncomplicated people—dealing with the many other guests was a punishing ordeal, 'as always.'[32]

Einstein received an honorary doctorate the next day (March 8). The ceremony was filled with pomp, as well as an abundance of the obligatory rhetorical fireworks. The only exception, in Einstein's view, was Langwerth's long but content-rich speech, in which he recounted the history of German–Spanish relations—'typically German, no rhetorical flourishes.' Einstein then visited a technical school, where more speeches were in store for him, 'nothing but speeches, even though well-intentioned.' In his last lecture, at the Athenaeum, Einstein discussed the philosophical consequences of relativity, making the following salient points: There are no privileged reference systems; in other words, the natural laws are the same in all of them; and Euclidean geometry is valid only in the absence of gravitation.

Einstein's patience during the ceremonial events of the past several days did not go unrewarded: he spent the evening at the Kocherthalers' home, playing chamber music with a professor at the music conservatory who impressed him greatly, both as an artist and as a superb violinist.[33]

His academic and social obligations now largely behind him, Einstein was able to enjoy his remaining few days in Madrid as a typical tourist. Kuno and Lina Kocherthaler took Einstein and Elsa on an excursion to El Escorial, the magnificent residence of the kings of Spain, and to the mountains north of Madrid. Together with some of their friends, they spent a relaxed, enjoyable evening in a small, primitive dance hall. On another night, Einstein and Elsa dined with Ambassador Langwerth's family. Most importantly, however, Einstein had time to pay several visits to the Prado, where he savored the 'glorious works' of Velasquez, El Greco, Goya, Raphael, and Fra Angelico.

On Monday, March 12, Einstein made his last entry in the travel diary. He and Elsa left Madrid for Saragossa on that day; he presented the first of his three scheduled lectures the same evening. According to contemporary accounts, the lecture hall was completely packed, adorned by the presence of several pretty young ladies and distinguished personages. Only a scant minority in the audience comprehended Einstein's first lecture, on special relativity. When at the second lecture the audience had shrunk dramatically, Jerónimo Vecino, Einstein's host in Saragossa, canceled the third lecture altogether.[34]

Einstein and Elsa spent Tuesday morning touring the city, visiting the Cathedral of Pilar and the Aljafería, the magnificent palace of the Muslim kings of Aragon. It was then time for a luncheon banquet at the Mercantile Casino, an affair to which the Academy of Sciences had invited Saragossa's academic elite. Einstein listened to numerous speeches before responding to the accolades that were heaped upon him. Afterward the Einsteins dined at the home of the German consul, where another opportunity for Einstein to play the violin had been arranged. He was accompanied by the distinguished pianist Emil von Sauer, who happened to be visiting Saragossa at the same time—a rare opportunity for an out-of-practice amateur violinist![35]

The following day (March 14) Einstein lunched with von Sauer at his hotel, and then he and Elsa boarded the train to Barcelona. After almost six months "on the road," they were finally on their way home.

4.

South America (1925)

BACK AT HOME

Their long journey to the Far East and their sojourns in Palestine and Spain behind them, Einstein and Elsa returned to Berlin in the spring of 1923. Einstein had spent six months partly as a tourist, partly as an itinerant lecturer addressing large but largely uncomprehending audiences; he was now ready to resume his academic duties at the university and the academy. He did not conduct conventional university courses but gave two kinds of lectures: on the one hand, he discussed his own current work with advanced students, and on the other, he gave popular talks about relativity and other scientific topics to general audiences. Since these audiences were often dominated by tourists who came to see the famous man, and because Einstein did not enjoy seeing their blank expressions, a few minutes into these talks he would pause and invite anyone with no further interest in the subject to leave. Usually only eight or ten genuine seekers of enlightenment remained.[1]

The political climate in Germany had hardly improved during Einstein's absence, but he kept a lower profile after his return than he had before. The harsh French occupation of the Ruhr region continued, in the face of great popular resentment and acts of sabotage, and so also did the economic crisis and the rampant monetary inflation. In September 1923, the price of a loaf of bread in Berlin rose to 10 million marks. A kilogram of beef cost 76 million marks. In November of the same year, Hitler orchestrated the inept Beer Hall Putsch with the support of General Ludendorff, but it was quickly put down by the Bavarian police. Hitler was arrested, tried, and convicted of treason, but he received only a mild sentence. He used his eight months of confinement to write *Mein Kampf*, his diatribe against Jews, Slavs, and democracy, and which expounded on his plan for world domination. Two years later, Ludendorff's wartime partner, Marshal Hindenburg, sought entry into German politics, and when the news of Hindenburg's candidacy for president reached Einstein in Montevideo, he was filled with justifiable apprehension. Seven years later, President Hindenburg smoothed Hitler's path to power.

In science, the second half of the decade produced an avalanche of important discoveries that were to affect the future directions of science and technology profoundly. Among the most far-reaching was Louis de Broglie's hypothesis that just as light can be described in terms of quanta (photons), so can particles be described in terms of waves. His proposal was soon confirmed experimentally and inspired the discovery of quantum mechanics a year later. The new mechanics, long anticipated by Einstein, opened up the detailed understanding of the subatomic world for the first time. At the other end of the scale, Edwin Hubble employed the new Mount Wilson telescope to demonstrate that the universe contained a vast number of galaxies beside our Milky Way and that the universe was, moreover, expanding rapidly; and, finally, Millikan's discovery of cosmic rays revealed the existence of previously unknown, elementary particles.

As these dramatic developments in physics unfolded, Einstein continued his search for a unified field theory, yet without making any progress. On the other hand, he did make an important contribution to quantum physics by developing the so-called Bose–Einstein statistics, generally regarded as the last of his many outstanding discoveries.[2]

AT SEA AGAIN

Einstein had been invited to South America several years earlier, but it was not until 1925 that he agreed to a lecture tour of several weeks, to three countries—including an enticing sea voyage.[3] On this trip, he was to be accompanied not by Elsa, but by the younger of his two stepdaughters, Margot; she fell ill shortly before their departure, however, so Einstein ended up having to travel alone.

The sun was shining brightly when a group of friends bade him farewell at the railway station in Berlin,[4] but when he arrived in Hamburg, his port of embarkation, the sky was overcast. He was met at the station by Marie Robinow, the wife of an eminent Hamburg jurist,[5] and her son-in-law. They brought Einstein to their home, where he spent the afternoon playing Mozart sonatas with Mrs. Robinow, using a child's violin borrowed for the occasion.

He spent the night in a hotel. In the morning, the Robinows' 'fine and intelligent' son-in-law escorted him to his ship, the *Cap Polonio*, a newly refurbished luxury liner of the Hamburg South America Line.[6] Half an hour later, the ship cast off, and Einstein stood on deck and watched as she pulled away from the dock and steamed down the Elbe past the many warehouses and ships. While most of the passengers had recognized his 'visage,' they were still preoccupied with their farewells, and so happily they left him alone—for the moment. Once the ship was underway, he settled in his luxurious cabin, in which the engine vibrations of the 'gigantic ship' were imperceptible—a vast improvement over his quarters on the *Kitano Maru*, next to the engine room. 'Blessed tranquility.' Einstein mused that it was too bad Margot was not there; she would have enjoyed it.

The *Cap Polonio* arrived in the harbor of Boulogne the next day, where many additional passengers, mostly South Americans, joined the ship. Although the sky remained overcast, Einstein felt that the air was already becoming more caressing—the exhilaration he had experienced on his earlier sea voyages was evidently returning. Though he had been at sea for only one day, he commented to his diary, 'thank God, the trip still has a long time to go, but I already dread the arrival.'

On the first day at sea, Einstein was fortunate to make the acquaintance of a fellow academic, Carl Jesinghaus.[7] The two men promptly arranged to be table partners. Jesinghaus, a philosopher and psychologist, was quiet and knowledgeable, and at their first dinner, the topic of conversation was causality; between meals, however, Einstein kept largely to himself. He read Chaucer's *Canterbury Tales* and enjoyed them greatly. He also worked on an idea concerning the foundations of Riemannian geometry.

In the morning of March 8, the ship approached Bilbao, on Spain's northern coast, and encountered brilliant sunshine for the first time. Einstein was delighted. The *Cap Polonio* was the largest ship to visit Bilbao, so that her arrival attracted a crowd of curious Spaniards, many of whom came on board. Einstein observed that these Spaniards were not at all blasé, but rather curious, childlike, and self-conscious, while the Spanish ladies, with their black eyes and black hair, wore little lace kerchiefs on their heads. Later that day, Einstein made the acquaintance of another congenial fellow passenger, Else Jerusalem, a feminist author of some renown. Einstein thought she had the elemental force of a 'panther cat'—and that is the epithet he gave her.[8] That evening, Einstein wandered into a room reserved for women and their children, and he stayed to play the piano for them—until this lovely scene was interrupted by the arrival of two Swabian ladies. They had dropped in, quite innocently, just wanting to listen, but they instantly put Einstein to flight. 'I must have cut a funny figure making my getaway,' he chuckled to his diary.

The next day, the *Cap Polonio* docked briefly in La Coruna, and then in Vigo, both ports near Spain's northwest corner. After discussing logic with Jesinghaus in the morning, he chatted with the 'Panther Cat,' concluding that while she was honest, impertinent, and vain, she was feminine only with regard to the last attribute. Still, the two seem to have gotten on well together. Einstein was often in her company, both on board and later, in Buenos Aires. Einstein spent most of the day on deck, relishing the 'intoxicating Southern sun.' When the ship entered the Bay of Vigo, surrounded by terraced hillsides on three sides, he felt as if he were in the nest of some giant bird.

Once arrived in Lisbon, on March 11, Einstein joined Jesinghaus and another passenger on a sightseeing stroll on shore. The city appeared quite dilapidated to Einstein, but it felt congenial all the same. Life in Lisbon seemed to be easygoing and unhurried, devoid of goals or awareness—so different from Berlin. Everywhere Einstein looked, it impressed upon him what life was like in ancient times. In one of the city's markets, Einstein was touched by the grace with which a woman fishmonger carried a platter of fish on her head— and made a proud, impish gesture to him, when he photographed her.

The three tourists then walked up to the Royal Castle, which afforded an imposing vista of town and harbor, before hiring a car, which took them along the coast to a monastery, whose playful, late Gothic architecture delighted Einstein. He particularly enjoyed the marvelous two-story-high-cloisters and the droll fountain lion. After returning to the ship, Einstein confessed to his diary: this 'raggedy land instilled in me a kind of longing.'

After leaving Lisbon, the *Cap Polonio* steamed south, following the coast of Africa for a few days while the weather grew noticeably warmer. When the ship passed Tenerife, the sun was setting and the steep green slopes of the mountain dominating the island were bathed in a brilliant light. In his cabin, Einstein continued to work on a unified field theory each day, without making progress. He turned to the books he had brought along and immersed himself in the latest philosophical works of Émile Meyerson and David Koigen.[9]

Einstein and Else Jerusalem dined with the captain on March 13. Their meal lasted three hours, with lively conversations that ranged over both serious topics and humorous ones. Einstein had come to like the captain and his sense of humor, and Else Jerusalem also grew in his estimation: she was both a Jew and an East German, 'both formidable specimens!'

The next day was Einstein's birthday; he was genuinely moved when he received a birthday card. He had his violin with him, as on most of his travels, and had managed to find a few string players among his fellow passengers—one was a young widow, and another was a student—but their music session was only moderately successful. That afternoon, Einstein and the captain listened as Else Jerusalem read her latest drama, which portrayed a young man torn between his Jewish background and a wider perspective and range of activity. A gripping, Jesus-like character, thought Einstein, but somewhat too abstract. Afterward, he tried to explain the basic ideas of relativity to Jerusalem, and in the evening, he and Jesinghaus discussed the essence of religion.

When the *Cap Polonio* passed the islands of Cape Verde, the precipitous flanks of the volcanic Fogo Island were brightly illuminated by the setting sun, while everything else was bathed in a subdued blue light. With the ship approaching the equator, and the weather growing increasingly tropical, 'King Neptune' appeared on deck and performed the traditional baptism of passengers crossing the equator for the first time: 'Harmless fun for everyone,' remarked Einstein.

The best part of Einstein's day was the daily string quartet session, for which he had now recruited 'the merry widow,' a certain Mr. Hollander, and the first violinist of the ship's orchestra. The group played Mozart and Schubert quartets in Einstein's steamy cabin, and 'although you sweat a lot, it is delightful.' The time for the ship's farewell party was approaching rapidly, and because all passengers were expected to perform in some way for the occasion, Einstein and his quartet partners rehearsed Mozart's *Eine Kleine Nachtmusik* in preparation for that gala event. Einstein was also prevailed upon to give a lecture on relativity to the ship's officers, and he commented plaintively to his diary that with the voyage drawing to a close, his splendid isolation was crumbling. Again he lamented that his stepdaughter was not there; 'it is nice to be alone, but not alone among a lot of strange monkeys.'

At the concert for first-class passengers on March 19, Einstein played with his string quartet and also performed Beethoven's Romance in F Major.

But the 'unspeakably stupid productions' presented by the Argentinian passengers appalled him: the performers evidently belonged to the wealthy classes, and while they were quite blasé, they were, at the same time, childish and naïve.

Jesinghaus introduced Einstein to some popular Argentinian music, said to stem from the Incas, and this set Einstein wondering what musical riches must have vanished along with native peoples. He had many conversations with the 'panther cat' in which she tried ceaselessly to fathom Einstein's persona, while he, in turn, had fun teasing her. He was amused by her manner, which he described as 'serious and brazen,' that of a typical Russian Jewish woman.

On March 22, the *Cap Polonio* reached Rio de Janeiro, where she would stay for a few hours before heading for Buenos Aires. Although it was raining lightly, the 'bizarre gigantic rock' (Sugarloaf Mountain) was visible, and its majesty impressed Einstein. A rabbi and a delegation of engineers and physicians welcomed him and offered to take him on a tour of the city while his ship remained in port. He accepted, and his very congenial guides then took Einstein to the botanical gardens, where he was astonished by the incredibly vigorous plant life. It seemed to him that everything was growing under his very eyes: it 'surpasses the dreams 1,001 Nights!' The mix of people of different ancestry—Portuguese, Indian, black, and everything in between—that he saw in the streets of Rio delighted him, and by the time he got back to the ship, he had collected an 'indescribable wealth of impressions' in just a few hours.

ARGENTINA[10]

Two days later, the *Cap Polonio* docked briefly in Montevideo, where another reception committee of journalists and 'diverse Jews' came aboard to accompany Einstein on the short passage to the ship's final destination, Buenos Aires. The group was headed by Mauricio Nirenstein, secretary of the University of Buenos Aires, a 'resigned and decent' man, according to Einstein, who was his guide while in Buenos Aires. He counseled Einstein on how to avoid the various political and philosophical controversies that were currently swirling in Argentina.[11] The other members of the welcoming group were, however, more or less 'unclean' (*unsauber*), according to Einstein. Eventually he had to be rescued from that 'unappetizing riffraff' by the stewards and Else Jerusalem. Apart from the commotion they caused, the ship encountered several other delays and did not dock in Buenos Aires until early the next day.

Early in the morning, on March 24, the hullabaloo created by the reporters and greeters on board started up again, but with Nirenstein's help Einstein made his escape and came ashore at 8:30 a.m. A photograph of Einstein as he disembarked shows him walking down the gangplank, his face unusually tense, surrounded by broadly grinning, excited greeters. He was quite exhausted by the time he arrived at the palatial home of Bruno Wassermann, a wealthy German Jewish merchant, where he would stay while in Buenos Aires. There he found peace and quiet, at last.

His hostess, the 'cheerful Senora Wassermann,' volunteered to act as Einstein's private secretary; her newly arrived friend Else Jerusalem acted as his interpreter in his encounters with the press. In the afternoon, several other ladies belonging to their circle arrived at the Wassermann home, as did the German ambassador. He was followed by Leopoldo Lugones, Argentina's most celebrated poet, author, and politician, as well as by representatives of several Jewish organizations.[12] In the evening, Einstein made the requisite courtesy calls on the president and the dean of the university: modest, straightforward, sober, and friendly people, but without a sense of mission. They reminded him, in some regards, of the Swiss and of other republicans. It seemed to Einstein that the city was comfortable but boring, and that its inhabitants were fragile, delicate, and one-dimensional. Coming from Germany, a country still reeling from the shock of mass slaughter, defeat, and political turmoil, Einstein saw luxury and superficiality everywhere he looked.

On March 26, the social and academic merry-go-round that Einstein had dreaded for so long began in earnest. He faced a horde of journalists and photographers in the morning before being taken on an automobile tour of the city ending in the Abasto, the city's largest food market. After calling on the dean of the university and his wife, Einstein met with a delegation of Jews, who invited him to take part in a celebratory mass meeting; but recalling his "belly-full" of mass meetings in New York, Einstein declined resolutely.[13] After conferring with a few more delegations, he was glad to spend a quiet evening at the Wassermann home; Else Jerusalem was there also, displaying her wit and high spirits. Einstein thought that her good cheer seemed forced, however.

The formal welcoming ceremony at the university took place the next day. After several ardent introductions, Einstein presented a brief, general lecture, speaking in stumbling French to a mostly unruly audience—all in all, it was 'an uncivilized occasion.'

The first of his several lectures at the university took place the next day, but not before he had listened to more flamboyant speeches by the academics and politicians gathered on the rostrum. The hall was overcrowded and sizzling hot. On this occasion at least, there were many young people in the audience who took an interest in what he said, which put Einstein in a much better frame of mind. Several 'inconsequential visits' followed, but he found them bearable. That evening he was a guest at a small dinner party in the home of the wealthy Alfredo Hirsch and his wife, a 'beautiful Jewish woman.'[14] Their luxurious home was filled with magnificent works of art, and there was even a pipe organ. Einstein mused that the urge to possess beautiful things was a barbarian's first step toward respectability; it reminded him of a child who is not satisfied with seeing a butterfly but needs to touch it or even put it in his mouth.

The next day was a rainy Sunday. Einstein spent the morning alone in his room, treasuring the blissful peace and quiet. He ruminated that a person has to be exposed to a great deal of turmoil before finding happiness in peace and quiet, and the past few days had prepared him sufficiently to attain that happy state. In the afternoon, he left his room and went for a walk with his host, Bruno Wassermann.

On Monday, March 30, Einstein delivered his second university lecture and then spent a pleasant evening with his cousin, Robert Koch, who resided in Buenos Aires. Since the two men were the same age, and had attended the cantonal school in Aarau at the same time, they had a lot in common and knew each other well. On this occasion, they marveled: 'how old we have become!' (Einstein was forty-six.)

The next day, Einstein was invited to inspect the plant of *La Prensa*, Argentina's largest newspaper, where he viewed the very latest automated printing presses. The enormous expenditure of paper and of human effort dismayed him; he asked if it would soon be time for the automated reader, as well. A visit to the city's Jewish quarter was on the agenda the following day. He was shown a newspaper office, an orphanage, and several synagogues (*shuls*), but the visit left Einstein unimpressed. It was a tragedy, he opined, that 'the Jewish people lost their soul along with the lice'; but was that not the case for other nations, as well?

On April 1, Einstein and señora Wassermann were taken on a sightseeing flight in a German Junkers seaplane, which happened to be visiting Buenos Aires. It was Einstein's first ride in an airplane, and he was duly impressed, particularly by the takeoff. In the afternoon, he was received by Argentina's president, Marcelo de Alvear, and several ministers; afterward, he was taken on a tour of the Ethnological Museum until it was time to deliver his third relativity lecture at the university. Later, Lugones brought Einstein to his home, where the two men spent the rest of the evening. It had been a busy day. Einstein ended his account of the day's activities with 'That'll do' (*Das reicht*).

The following day, Einstein traveled to the town of La Plata, forty miles from Buenos Aires. The quiet, pretty town charmed him, reminding him of towns in Italy. He was obliged to attend the formal start of the new term at the university, which was housed in superb, American-style buildings. The drawn-out academic ceremony included a number of exceedingly long speeches and musical performances.

Although there is no mention of him in Einstein's diary, his host in La Plata was Richard Gans, the director and founder of the university's physics institute. Gans was an accomplished German Jewish physicist who had immigrated to Argentina in 1911 but disapproved of Einstein's political views. Indeed, Gans harbored such fervent nationalist sympathies that he returned to Germany soon after Einstein's visit. His subsequent history in Hitler's Germany is most surprising and constitutes a fascinating story.[15]

Back in Buenos Aires, Einstein gave his fourth relativity lecture on April 3, afterward dining with the rector of the university. In a talk to the philosophy faculty the next day, he discussed ways of conceptualizing spherical space. He was then finally able to relax, spending another serene evening at the home of his cousin, Robert Koch, whose uncle, the French industrialist Louis Dreyfus, was also present. Einstein was impressed by Dreyfus, finding him very intelligent, canny, and good-natured.

On Sunday (April 5), the Wassermanns drove with Einstein to their country estate of Lavallol, a place where he was able to relax and escape the commotion surrounding him in the city. Lavallol was his sanctuary on this and several other occasions.

Back in Buenos Aires, Einstein visited a laboratory where experiments had demonstrated that exposure to intense monochromatic light caused subjective apparitions to appear on one's retina—this is an example of the 'inconsequential visits' his hosts sometimes arranged. There was a large

Zionist meeting in support of the Hebrew University in the evening, and after Einstein had listened to a number of pathos-laden speeches by the 'Spaniards,' he gave a short speech of his own. Benzion Mossinsohn, whom Einstein had last seen on his visit to Palestine in 1923, gave a homespun speech in Yiddish, a language that Einstein found to be remarkably warm-hearted and expressive.

The next day, Einstein met with the eminent physician and the rector of Buenos Aires University, José Arce.[16] Einstein was sufficiently impressed by Arce's clinic, as well as by the man, to comment on how much he stood out from his surroundings. Fortunately, Einstein was again able to recuperate from his crowded schedule during a three-day stay at Lavallol, the Wasser-manns' estate. While he was there, he had a 'marvelous idea for a new theory' to connect electricity and gravitation.

On April 11, Einstein boarded a special railway carriage for the overnight trip to Córdoba, a city some four hundred miles northeast of Buenos Aires. He was not alone; with him traveled the philosopher Cori-olano Alberini, Mauricio Nirenstein, Enrique Butty, and several others.[17] On their arrival in the morning, Einstein was taken on a sightseeing drive through the barren and forbidding granite mountains west of the city. The excursion over, he was honored at a 'very boring' government dinner.

In the morning, Einstein delivered his lecture—he would give only one at Córdoba—in a magnificent hall at the university and was then celebrated at a festive academic assembly. He sat next to the newly elected provincial governor, 'a very fine and interesting person,' but apart from him and Alberini, Einstein had had his fill of the 'tiresome profusion of Spaniards, journalists and Jews.' It took a droll speech delivered in Hebrew by a 'trem-bling virgin' to restore his good humor.

Einstein admired Córdoba's wonderful cathedral, and he appreciated the well-proportioned houses without any silly decorations. He thought that he could detect remnants of the old Spanish culture, with its love of the land and awareness of higher things—at the price of being ruled by priests (*Pfaffen-herrschaft*). Einstein decided that even that was preferable to a self-satisfied civilization devoid of culture.

On April 14, Einstein traveled back to Buenos Aires, overnight, in the same private railway carriage in which he had come. He was glad to be back, but he was in a sour mood and a terribly misanthropic frame of mind (*bin schrecklich menschenmüde*). Most of the people he encountered struck him as 'lacquered cigar-store Indians, skeptical-cynical, indifferent to culture, and debauched in oxen fat.'[18] He felt severely oppressed by the thought of having to roam about there so much longer.

Two days later, Einstein met with the executive committee of the Zionist organization in the morning and was honored at a session of the National Academy of Exact Sciences in the afternoon. After he was elected as an external member, the academicians asked him such remarkably stupid scientific questions that it was difficult for him to keep a straight face. He was photographed by a portrait painter the next day, and later the same day, he gave the penultimate university lecture of his visit. Einstein's day was not yet over: he was honored at an evening reception given by the German embassy and to which only Argentinian dignitaries had been invited—no Germans were among the guests. Einstein suspected that this was due to the hostility that members of the large German community felt toward his pacifism. 'A droll lot, these Germans,' mused Einstein, 'I am a foul-smelling flower to them, but they keep sticking me in their buttonhole, all the same.'

Einstein's suspicions regarding the guest list for the embassy reception were confirmed by the report sent by Karl Gneist, the German ambassador, to the Foreign Office. He gave a glowing account of the popular interest generated by Einstein's visit: "Hardly a day went by without the papers bringing many columns of stories related to the person of the scientist and his theory. . . . The local German community, unfortunately, stayed away from all events because some members had objected to comments Einstein made in an interview with Nación, as being pacifist. . . . For the first time, a world-famous German scholar came here, and his naive, kindly, perhaps somewhat unworldly manner had an extraordinary appeal for the local population. One could not find a better man to counter the hostile propaganda of lies, and to destroy the fable of German barbarism."[19]

On April 18, at the Wassermanns' home, Einstein gave a private lecture to señora Wassermann's circle of female friends, but the Panther Cat was conspicuously absent. She was evidently miffed at Einstein for his having neglected her.[20] Later, he addressed the Societa Hebraica, and in the same lecture he discussed both the spirit of Zionism and the size of atoms. The next day being Sunday, he relaxed at the country house in Lavallol, but in the evening there was yet another reception of Jewish Societies, with a speech by Mossinsohn and with singing. At the last university lecture (April 20), he spoke before a particularly enthusiastic audience, but on the following day he attended 'a very tasteless reception' at a Jewish hospital. He found that occasion so annoying that he gave the organizers a dressing-down.

After five weeks in Argentina, the time had come for Einstein's farewell. At a luncheon for his closest associates in Buenos Aires, he presented photographs of himself to señora Wassermann and Professor Nirenstein, each inscribed with an appropriate, witty poem. Einstein had also prepared such a memento for Else Jerusalem, but the 'Panther Cat,' still miffed at Einstein, was again missing from the circle.[21]

The official farewell breakfast on April 22 was attended by all of the scientific and political bigwigs.[22] A more informal farewell party, given by the university students, took place in the evening—this was more to Einstein's taste. There was a lot of singing and guitar playing, and to cap it off, Einstein played his violin.

URUGUAY

The trip to Montevideo, Einstein's next destination, was made on the overnight steam train *City of Buenos Aires*, arriving on April 24. Einstein's lectures and speeches in Argentina had been widely reported in the Uruguayan press. As a result, an enthusiastic crowd had gathered at the station to welcome him. His hosts took him first to the top of a tall insurance company building to enjoy the panoramic view of the city below. They offered him a choice of accommodations: he could accept the city's official invitation to reside at the elegant Parque Hotel or be the houseguest of Naum Rosenblatt and his Russian Jewish family. Einstein chose the latter.

When he was settled in the Rosenblatts' home, the German ambassador, Arthur Schmidt-Elskop, came to call, followed by the philosopher Carlos Vaz Ferreira, who took Einstein on a leisurely stroll through the city.[23] In his diary, Einstein described Vaz Ferreira as a fine, nervous black chap, whose French—their only common language—was even worse than his own, adding, 'He was in awe of me, as are most people.'

Einstein felt palpably more at ease in Uruguay than in Argentina. He quickly took a liking to the 'warm-hearted and guileless' Rosenblatt couple, who spoke only Yiddish, and to their children, with whom he spoke in French. He also felt at ease in the company of the engineering professor Maggiolo, 'a fine person, quiet and introverted, not at all American,' and with Armando Castro, who was another engineer, and his 'charming, red-cheeked little son.'

On the day after his arrival, Einstein gave the first of his scheduled lectures on relativity, which was followed by a solemn reception at the university. In the evening, Maggiolo, the Rosenblatts, and the German ambassador took Einstein to a performance of *La Traviata* that was put on by a visiting Italian troupe; on the following evening, the company gave a performance of *Lohengrin* especially for Einstein's benefit. Their presentation alternated between good and comical—and not only on account of the troupe of performers, thought Einstein. Two students were assigned to stand watch and to protect Einstein's privacy at all times. A touchingly obliging manservant was assigned to him, with whom he could communicate only in sign language.

After only two days in the country, Einstein came to the conclusion that Uruguay was a fortunate little country. Here he found genuine warmth and love of the land, without megalomania. The country had a pleasant, warm, humid climate and exemplary social institutions, including exceptional care for children and the elderly, protection of illegitimate children, and an eight-hour workday. The country's constitution was, moreover, very liberal—similar to that of Switzerland—and with strict separation between church and state.

On April 27, Juan Antonio Buero, the president of the senate, 'a polished and clever young man,' along with Maggiolo, Castro, and some others, took Einstein to a factory where very beautiful marble was produced. In the afternoon, Einstein called on the president of the republic, the education minister, and the local Swiss consul (the consul had attended the same school in Aarau as he had). Later, Einstein delivered the second lecture of his visit, and then spent a quiet evening with the Rosenblatt family.

Montevideo's German community honored Einstein at a genial reception the next day, 'with coffee accompaniment.' Most likely only the most liberal members of the community had been invited, even though the German ambassador, Schmidt-Elskop, reported Einstein's visit to the Foreign Office in Berlin in glowing terms: "Einstein left a superb impression on account of his very simple and sympathetic manner. Since he is celebrated everywhere as the 'sabio alemán' [German scholar], his visit has been very useful for the German cause."[24] In the evening, Jewish groups and groups supporting the League of Nations honored Einstein at a banquet. He was fortunate in being seated next to an interesting Englishman who had worked alongside Fridtjof Nansen, the distinguished humanist and explorer.[25]

Einstein gave his last lecture the following day (April 29), and in the evening, the German ambassador gave a large reception at which Uruguayan politicians and scholars were the only guests. Evidently in this case too, Einstein's political views were too controversial for many resident Germans.

The next day was Einstein's last full day in Montevideo. He attended a special screening of two films: a documentary film about the recent expedition to the South Pole, and a charming Chaplin film. In the afternoon, Einstein was taken sailing, which he enjoyed enormously; finally, at 9:00 p.m., there was another reception and banquet, sponsored jointly by the government and the university. Einstein was seated between the president and a minister, and he had a thoroughly good time. In the course of the ceremony, the orchestra struck up "Die Wacht am Rhein" ("Watch on the Rhine") instead of the German national anthem, and Einstein exchanged sly grins with Schmidt-Elskop, for they knew the song as the rallying cry of militant nationalists.

After his stay in Montevideo had drawn to a close, Einstein felt the need to summarize his impressions while there. Everyone had been so kind and unceremonious—even though he had been obliged to wear the hated tuxedo. He felt that his brief diary entries had not done justice to the reality, that his experiences had been much richer and more varied, so much so that often he could hardly catch his breath. He had enjoyed his stay in Uruguay more than his time in Argentina, because he had found a greater feeling for humanity there, which was, of course, partly due to the smaller size of the country and the city. The people in Uruguay seemed modest and natural—they reminded Einstein of the Swiss and the Dutch: 'The devil fetch the large nations and their manias, if it were up to me, I would cut them all up into smaller ones!'

As chance would have it, Einstein was scheduled to leave for Rio de Janeiro by sea on May 1, Labor Day, a public holiday that was taken very seriously in Montevideo. All work stopped, and automobiles were not allowed on the city streets. The French liner *Valdivia*, on which Einstein was to travel to Rio, was at anchor in the harbor. An official car and a harbor launch had to be found to transport him and the large contingent of farewell-wishers aboard.

The *Valdivia* was small and very filthy, Einstein discovered.[26] He slept quite poorly in her, and her toilet facilities appalled him—but they were endurable for three days. Her crew, on the other hand, were very friendly. And although he deplored the heavy, poorly prepared food, Einstein liked the captain. He appreciated the relaxed atmosphere at the captain's table and also his sensitive and unobtrusive table companions, who he felt were more agreeable, more modest and natural, than Germans.

In spite of the *Valdivia's* obvious shortcomings, Einstein savored his three restful days aboard and used the opportunity to reflect on his South American expedition. Still, he was filled with apprehension, knowing that the blissful calm of shipboard life would end as soon as he arrived in Brazil. He would give a lot if he did not have to 'get on his trapeze once more in Rio,' but with God's help, he expected that he would be able to tolerate a few more days of this monkey business (*Affenkomödie*)—and afterward, he could look forward to a magnificent, long sea voyage home. He had so much turmoil and so many changes in scenery behind him that he could scarcely imagine leading a normal, quiet life again.

Einstein's discovery that all the scientific ideas he had developed in Argentina turned out to be useless hardly improved his mood. He was, moreover, deeply distressed by the news that reached him from Germany, particularly the report that Field Marshal Hindenburg had entered politics again. His nomination in the presidential election was an embarrassment for the German ambassador in Montevideo, as it gave Uruguayans occasion to make fun of the Germans. Germany was a nation, commented Einstein, 'that had its good sense beaten out of it with a stick.'

BRAZIL[27]

On May 4, at sunset, the *Valdivia* dropped anchor in the harbor of Rio de Janeiro among its fantastically shaped granite islands. The weather was excellent. Taken ashore, Einstein was welcomed by the waiting 'professors and Jews.' He quickly gained the impression that they had all been softened by the tropical climate, musing that a European required greater stimulation of his metabolism than this eternally warm and humid ambiance could provide. Even the life of a European wage slave was richer and, above all, less dreamlike and hazy than theirs. Einstein's ideas regarding the debilitating effects of climate are reminiscent of his view that it was the climate that accounted for the abject poverty he had witnessed in the Far East. They reflect popular nineteenth-century theories that the climate of the New World had an incapacitating effect on Europeans.[28]

Einstein took up residence in the luxurious Hotel Glória, situated on Rio's waterfront. In the morning, he was visited by Isidoro Kohn, an Austrian-born businessman who was the president of Rio's Jewish community—a bright and agreeable gentleman, although also something of a busybody (*Geschaftlhuber*). Einstein joined Kohn on a walking tour of the city, and for lunch they were joined by Kohn's wife and another woman, two lighthearted ladies who were appealing table companions for Einstein. In the afternoon, he met with German businessmen, and then 'the professors' took him on the dizzying cable-car ride, high above a wild forest, to the top of Sugarloaf Mountain. Einstein admired the spectacular view and the superb interplay between fog and sunlight, something that often intrigued him. Later came a reception hosted by Jewish groups, and afterward, Einstein enjoyed a nocturnal automobile ride with the 'fine and intelligent' Rabbi Isaias Raffalovich, who was the spiritual leader of Rio's Jewish community and had played a major role in bringing Einstein to Brazil.

The following morning, Antonio da Silva Mello, a physician who had reformed Brazilian medicine according to the German model, accompanied Einstein on a stroll through Rio's upper town. During their walk, Mello tried to forewarn Einstein about the numerous petty intrigues in which the local academics were embroiled. After lunch in a harbor tavern, in which Einstein consumed a spicy fish dish, Einstein made the obligatory courtesy calls on Brazil's president, Artur Bernardes; on the minister of education; and on the mayor of Rio.

It was then time to head to the Engineering Club, where Einstein delivered the first of his scheduled lectures on relativity. Since there were very few physicists in Brazil at the time, Einstein addressed himself to a general audience—as he did in most of the lectures on this tour. The lecture hall was again far too crowded, as many officials in the audience had brought their wives and children along. Because of the tremendous heat, all the windows of the hall had been opened, and the noise from the street made it impossible to understand Einstein's words—for purely acoustical reasons. He decided that from a scientific point of view, his talk made no sense whatever, and the audience simply regarded him as a sort of 'white elephant,' while he, in turn, perceived them as fools (*Affen*). At night, back in his hotel room, he sat naked in front of the window, enjoying the moonlit panorama before him—the islands in the bay, some rocky, others clothed in green foliage.

The next day (May 5), Einstein visited the natural history museum, which was largely devoted to zoological and anthropological exhibits. He was fascinated by the beautiful construction of a snake's spinal column, but he was above all interested in the exhibits of native culture. He examined various artifacts, including shrunken mummies and poison arrows used by the Amazonian Ticuna Indians. To Einstein's regret, there was not enough time to hear recordings of native Indian music.

Dr. Mello brought Einstein to the home of Professor Aloysio de Castro for lunch. While Einstein was unimpressed by his host, he enjoyed his other table companions: a Russian archeologist, an intelligent journalist, and, particularly, a pretty, intelligent, and somewhat arrogant author.[29] That afternoon, at the Academy of Sciences, he had to sit through more bombastic speeches, given in Portuguese, before delivering his own talk in French. Einstein commented in his diary that his Brazilian hosts were truly prodigious orators and that they lauded a person by extolling his eloquence. Einstein felt certain that all this tomfoolery and these irrelevances (*Unsachlichkeit*) must have something to do with the climate—but acknowledged that nobody else agreed with him.

It is noteworthy that when Einstein addressed the members of the academy, he did not speak of relativity as usual but instead reviewed the present state of knowledge of the nature of light—a topic that had occupied him intensely for a long time. Twenty years earlier, he had explained the photoelectric effect by proposing that light travels as quanta of radiation (photons), even though this corpuscular model flew in the face of a great deal of evidence (e.g., interference, diffraction, refraction) that light was an electromagnetic wave. These conflicting aspects of light troubled Einstein profoundly, and he had often urged physicists to search for a new mechanics that could resolve the dual nature of light. Shortly before he left on his South American tour, the issue had come to a head when the experiments of Arthur Compton (see chapter 7) provided strong evidence for the corpuscular nature of photons. A new theory (the BKS theory) had attempted to explain Compton's data on a purely statistical basis. Speaking now at the academy, Einstein provided the background to the dichotomy of light and announced that an experiment was even then under way in Berlin that would decide between the corpuscular and statistical explanations of Compton's data. The preliminary experimental results, which he had seen before he left on his journey, favored the corpuscular point of view, but definitive results would be ready when he returned to Berlin. The text of Einstein's talk was fortunately preserved, and it represents a lucid summary of the ambiguities of quantum physics only months before quantum mechanics was able to resolve them.[30]

On a tour of the Oswaldo Cruz Institute (for biomedical research) the next morning, Einstein used a microscope to observe trypanosomes, the tropical parasites responsible for Chagas' disease. He was guided by Carlos Chagas, the discoverer of the life cycle of trypanosomes, for whom the disease is named. In the afternoon, Einstein delivered his second lecture at the Engineering Club—as before, in stifling heat and in an overcrowded lecture hall. His evening was, however, very agreeable. He spent it at "Germania," the German club, where he dined with members of Rio's German community—and did not have to speak in French. In his report to the Foreign Office, the German ambassador, Hubert Knipping, remarked with evident relief that the event had come off harmoniously: "Einstein's modest nature earned him a great deal of personal sympathy, which was not diminished by his indifference in sartorial matters, as was abundantly in evidence. His visit has undoubtedly benefitted the German cause here."[31] The sartorial indifference refers to Einstein's careless approach to matters of grooming, which had become mythological in the local press—presumably Elsa, who was not with him on this trip, was accustomed to managing his appearance.

The following day (May 9), Einstein visited the National Observatory and met its director, Henrique Morize. During the 1919 solar eclipse, Morize had participated in Eddington's expedition, whose observations confirmed a critical prediction of general relativity theory. Einstein, ever interested in ingenious scientific instrumentation, was shown the new, ultrasensitive Milne-Shaw seismograph, the first he had seen. At lunch, at Mello's home, he enjoyed several tasty Brazilian dishes, and afterward he was taken to see two physiologists who showed him the results of their investigation of respiration. In the evening, Einstein dined at the home of Isidoro Kohn and his family, whom he described as 'ordinary, but well-meaning people.' His day was not done, however, for at nine o'clock, he was feted at a large reception of Jewish organizations at the Jockey Club. Inevitably, the occasion produced excited and long-winded speeches with 'lots of extravagant flattery,' even though well intentioned. 'Thank the Lord, that's done with,' he sighed to his diary later that night.

Einstein was now nearing the end of the patience and fortitude he had displayed throughout the tour. He confessed to an almost irresistible longing to get away from all these people he did not know. He consoled himself with the thought that there were just two more days to get through, and that they promised to be agreeable ones.

Einstein's penultimate day in Brazil did turn out to be very enjoyable. He joined the Kohn family and several others on an outstanding automobile excursion to a number of panoramic sites in the vicinity of Rio. At sunset the party rode the cogwheel railroad to the top of the Corcovado, the enormous granite peak that offers spectacular views of Rio and of Sugarloaf Mountain. (Corcovado was not yet surmounted by the famous statue of Jesus with outstretched arms.) Later that evening, Einstein attended a reception at the Zionist headquarters, which was held in a crowded and sizzling-hot room without any perceptible ventilation, but the speeches were, for once, mercifully short.

He visited the National Psychiatric Hospital the next morning and met its medical director, Juliano Moreira, who was of mixed race and, in Einstein's view, was an exceptionally brilliant person. Moreira had been responsible for reforming Brazilian psychiatry along lines he had observed while studying in Germany. Einstein lunched at Moreira's home, with 'lots of pepper and a German wife.' That afternoon, he made several obligatory courtesy calls on ministers of state, many of whom were absent, 'thank the Lord.' When that was over, he submitted to one of those abhorrent photography sessions and was then taken to a showing of a documentary film about the life of native Indians and about the humanitarian work done by General Rondon on their behalf.[32] (Einstein was so impressed by Rondon's leadership and the humane approach he used to prepare Indians for life in the modern world that after his return to Berlin he nominated him for a Nobel Peace Prize.) That left only two more functions he had to attend: a discussion at the Brazilian Press Association, and a dinner in his honor at the Hotel Glória given by Ambassador Knipping.

The next day (May 12), Einstein bade his farewells to his Brazilian hosts and embarked on the voyage home that he had looked forward to so longingly. His diary entries come to a precipitous end on a plaintive note: 'Free at last, but more dead than alive.'[33]

BACK IN BERLIN

The *Cap Norte*, which brought Einstein back from South America, was smaller and slower than the *Cap Polonio*, which had taken him there.[34] Once aboard, and for the first time, he forsook the travel diary completely. The lecture tour had been long and far more strenuous than expected, and even after spending two restful weeks at sea, Einstein was nervous and exhausted when he arrived at home. His doctor urged him to forego any similar undertakings for several years.

The misanthropic mood that pervades the last several diary entries suggests that the tour had not lived up to his expectations, but what were his expectations? They could hardly have been scientific in nature—there were very few theoretical physicists at the universities he visited, and most of his hosts had been physicians, philosophers, or engineers. Nor did Einstein have close ties to the various Jewish communities and Zionist groups that sponsored his visit. Most likely, the substantial honoraria he received, and an urge to escape the politics at home, played important roles in his decision to tour the three countries.

When he returned to Berlin in early June 1925, the political atmosphere in the city was as tense as it had been in March. In that year, Hitler published the first volume of *Mein Kampf* and took complete control of the Nazi Party as its *Führer*. Joseph Goebbels was appointed gauleiter of Berlin and instigated more violent confrontations with Hitler's political opponents, as well as attacks on Jews.

At the same time, Berlin's intellectual and artistic life was teeming with energy: Otto Dix, George Grosz, and other artists pursued a stark objectivity (*Sachlichkeit*) in displaying contemporary society; Gropius and the Bauhaus group gave architecture a new direction; Schoenberg rejected classical tonality in music; Brecht reinvented opera with the *Three-Penny Opera*; Brod published Kafka's *The Trial* posthumously; and Freudian psychoanalysis was spreading far and wide. Also in this brief golden period of the Weimar Republic, Heisenberg, Born, Schrödinger, and Dirac, among others, developed quantum mechanics—the culmination of quantum physics to which Einstein had made so many fundamental contributions.

Quantum mechanics made a detailed understanding of the subatomic world possible for the first time. Einstein admired its great successes, but could not love it—the probabilistic interpretation of the uncertainty principle remained indigestible to him.

DINNER WITH COUNT KESSLER

Einstein is often characterized as an intensely private person, a loner. But this is surely not the whole story, as witnessed by the remarkably active social lives he and Elsa led in Berlin. They were frequent visitors to theaters and concerts and were in great demand as dinner guests. This is borne out in the diary of Harry Graf Kessler. When Einstein and Elsa dined at Kessler's home on February 15, 1926, it was the first time Kessler saw Einstein since his return from South America. Kessler's dinner guests often included illustrious personages from the worlds of politics, journalism, the arts, and business, and occasionally they included members of the nobility. On this occasion, most of his guests were active supporters of the continuing efforts to encourage French–German reconciliation.[35] According to Kessler, Einstein made a majestic impression, in spite of his extreme modesty. He also reports that Einstein had gained a little weight and that he wore laced boots with his dinner jacket. Kessler added that Einstein had a mischievous look—his eyes had an almost childlike twinkle.

In the course of the evening, Elsa told Kessler that after admonishing her husband repeatedly, he had finally picked up the two gold medals he had been awarded by the Royal Astronomical Society. Later that afternoon, she met him at a cinema, and when she asked what the medals were like, he replied that he had not opened the package. Such knickknacks were of no interest to him, Elsa continued. She gave other examples of Einstein's disdain for honors and decorations: he absolutely refused to wear his decoration *Pour le Mérite*. Recently, at a meeting of the Prussian Academy, Nernst had pointed out to him that his wife must have forgotten to hang his *Pour le Mérite* around his neck, to which Einstein replied: "Not forgotten, no, not forgotten, I did not want to wear it." When the dinner conversation turned to the recent sensational discovery that Sirius ("the Dog Star," in the constellation Canis Major) was orbited by an enormously massive companion star, Sirius B, Einstein explained the significance of the discovery to the guests: Because Sirius B was enormously dense, general relativity theory predicted that its spectrum would be enormously redshifted. Kessler recognized and remarked on Einstein's ever-present slightly ironic facial expression, the "bemused and pained skepticism that plays around his eyes." It reminded him of someone who smiles at human conceit, not just superficially, but down to its very roots.[36]

5.

New York and Pasadena (1930–1931)

BERLIN 1925–1930

Einstein had barely recovered from his exhausting South American tour when he plunged right back into the spirited scientific, artistic, social, and political life of Berlin. Five years were to pass before he would undertake another voyage overseas. Some of the significant events in politics, science, and in Einstein's personal life that occurred in the interim are briefly summarized here.

In August 1925, two months after his return and while still recuperating, Einstein again took advantage of the Kiel "refuge" that his friend Hermann Anschütz had put at his disposal in the aftermath of Rathenau's murder (see chapter 2). Einstein referred to the apartment in Kiel as his "tub of Diogenes"—the legendary habitat of that philosopher in the streets of Athens. Einstein's tub was more luxurious, with a living room that opened onto a lawn and a river, where a sailboat was moored for his use. Occasionally, when Einstein was out sailing in the Bay of Kiel and became becalmed, an Anschütz launch was dispatched to rescue him.

On this occasion he had brought his elder son, Hans Albert, along for a sailing holiday. While the two were staying in Kiel, Einstein also worked on the design of the new gyrocompass that the Anschütz Company was preparing to manufacture. He joined Anschütz on board a German navy torpedo boat to witness sea trials of the gyrocompass—apparently, without perceiving any conflict with his pacifist convictions. For his contributions to the gyrocompass design, Einstein was to receive a royalty of 3 percent of its sale price.[1]

Anschütz was, however, much more than a business associate to Einstein. Later that year, Hans Albert announced his plan to marry Frieda Knecht and met with strenuous objections from both Einstein and Mileva. Anschütz offered to help out. He invited the two lovers to his home in Munich, where he attempted to change their minds. He offered Hans Albert an engineering job in Kiel and urged him to delay the marriage. But his efforts came to naught. Just as Einstein's family had failed to prevent Einstein from marrying Mileva twenty-five years earlier, so did Hans Albert's parents fail in their attempt. The pair was married in 1927. Einstein was soon reconciled with the couple, and unlike his own first marriage, theirs turned out to be a satisfying and long-lasting one.

Einstein suffered a far greater personal setback when Eduard, his younger son, was diagnosed with schizophrenia. He was eventually admitted to Burghölzli, a psychiatric institution in Zurich, where he received little medical relief, and where he spent the rest of his life. According to Elsa, Eduard's sorry fate affected Einstein much more deeply than he cared to show. He was convinced that the boy had inherited his illness from Mileva, whose sister was also a patient in a mental ward.

In the world of physics, those years were filled with exciting developments. In 1926, Schrödinger demonstrated that his and Heisenberg's formulations of quantum mechanics were equivalent, which opened the door wide to a new era in physics. Einstein applauded the success of the new mechanics, but the probabilistic interpretation of the "uncertainty principle" remained unpalatable to him. He devised numerous thought experiments to reveal inconsistencies in the uncertainty principle, but in each case, Bohr and others were able to show that the analysis was flawed. In spite of his philosophical reservations, Einstein nominated Schrödinger and Heisenberg for the Nobel Prize.

In March 1928, Einstein took part in a congress held in the little town of Zuoz, in Switzerland, to promote international understanding. He interrupted his stay there to travel briefly to Leipzig, where he appeared as an expert witness in a patent dispute between the Siemens and AEG corporations. Twenty years had passed since Einstein had worked in the patent office, but his interest in patent matters was evidently undiminished. He returned to Zuoz late at night, and after trudging uphill through deep snow, carrying his suitcase, he suffered a circulatory collapse that kept him confined to his bed for four months, although his full recovery took over a year.

While Einstein was convalescing at the seaside resort of Scharbeutz, on the Bay of Lübeck, he sent word to friends and colleagues that he had discovered a "distant parallelism" geometry that promised to open the door to a unified field theory. He submitted the proposed theory to the Academy of Sciences late in 1929. Even before it was published in the Proceedings of the Academy, reports of his new theory appeared in the press and generated enormous popular interest. However, knowledgeable physicists soon identified glaring flaws in it, and the illustrious Wolfgang Pauli wrote to Einstein that he deplored Einstein's decision to join the pure mathematicians and predicted that Einstein would renounce his new theory within a year. It actually took two years.[2]

While this strange episode puzzled fellow physicists, Einstein appeared unperturbed by it and without interruption resumed his quest for a unified field theory. The affair did not diminish his fame, and he continued to exploit his celebrity status to promote the two humanitarian causes closest to his heart: pacifism and the Hebrew University in Jerusalem. He justified his pacifist stance by contending that the welfare of humanity must take precedence over loyalty to one's country. But even like-minded people questioned his idealistic assertion that governments would be unable to wage wars if just 2 percent of the population refused to serve in the military.[3]

The year 1928 saw the Berlin premiere of Brecht and Weill's *Three-Penny Opera*. In the same year, amid the economic Depression, the number of unemployed workers in Germany reached the two million mark. This was also the year in which Einstein hired Helen Dukas as his secretary. She came from the same region in Swabia as Elsa and quickly became an indispensable member of the Einstein establishment, a position she would retain for the remainder of Einstein's life.

Einstein turned fifty in March 1929. Berlin's city council wished to mark the occasion by presenting the city's most famous citizen with a lakeside country home. Their laudable intentions became bogged down in a series of bureaucratic slip-ups and city politics, and in the end, Einstein declined the birthday present and built his own country home, a simple wooden structure in rural Caputh. The house was within easy reach of Berlin and was not equipped with a telephone. It quickly became Einstein's sanctuary from the demands of the city.

In the political arena, Germany was finally invited to join the League of Nations in 1926, and for a while, Einstein served on the League's commission on intellectual cooperation. However, he quickly became embroiled in fierce quarrels and resigned. Einstein's similarly brief tenure on the board of governors of the University of Jerusalem came to an end when his vision of who should be the first chancellor was at odds with the views of some influential financial backers.[4]

In September 1929, the dirigible *Graf Zeppelin* created great excitement when it circumnavigated the earth. A month later, the crash of the New York Stock Exchange was a harbinger of the ensuing worldwide Depression.

In the wake of the 1930 election, the Nazi Party became the second largest body in the Reichstag (German parliament), trailing only the Social Democrats. Yet Einstein continued to see militarism as a greater threat to peace than Hitler. During his upcoming American trip, he blamed Hitler's current success on the desperate, but transitory, economic situation in Germany.

At the sixth Solvay Conference of theoretical physicists, held in Brussels in October 1930, Einstein and Bohr continued their dialogue on the interpretation of quantum mechanics. Einstein also used the opportunity to pay another visit to the Belgian royal couple in their Brussels castle, where he played chamber music with the queen, Elizabeth.[5] In a letter to Elsa, Einstein described the simple vegetarian dinner he had shared with his royal hosts: spinach with a fried egg and potatoes. He added that no servants had been present and that he liked the couple enormously and was certain that the feeling was mutual.

Also in 1930, Arthur Fleming, a wealthy board member of the California Institute of Technology (Caltech), visited Einstein at his Caputh home and repeated Robert Millikan's invitation to spend two months as a research associate in Pasadena.[6] The visit would allow Einstein to interact with the noted physicists and astronomers at Caltech and at the Mount Wilson Observatory. Using the observatory's new 100-inch Hooker telescope, Edwin Hubble had demonstrated that the universe contained a vast number of galaxies, and that they were receding from each other; in other words, that our universe is expanding—a finding that would induce Einstein to modify his relativistic field equation after he returned.[7] An extended visit to Pasadena was clearly justified on scientific grounds; the generous salary offered by Fleming made Einstein's decision to accept the invitation even easier.

VOYAGE TO AMERICA

On November 30, 1930, Einstein and Elsa began their second journey to the United States at the railway station "Zoo," where they were seen off by friends and family members—along with the inevitable reporters and photographers. When Einstein at last went aboard, their sleeping car compartment was full of their luggage, but Elsa was missing. After the train left the station, Elsa was still nowhere to be seen and when she finally did appear, she was furious because she could not find their tickets, which turned out to be already with the conductor. The couple gradually calmed down after all the excitement of the departure and went to sleep. They arrived in Cologne at 8:00 a.m. and made a mad dash to their connecting train, which was waiting on another platform. Once on board, Elsa's fears that the porter had left behind a piece of luggage turned out to be unfounded.

Three hours later, they arrived in Liége (*Lüttich*) in bright sunshine. Einstein's uncle, Cesar Koch, was waiting at the station with one of his granddaughters. The four walked along the Maas River all the way to Cesar's home, exchanging family news and reminiscences. On arrival, Einstein was overjoyed to find his 'happily divorced' cousin Suzanna also present. He thoroughly enjoyed the patriarchal family meal that followed, for it had been prepared 'with love and masterly skill.' When the meal was over, a group of gentlemen arrived to discuss the political situation with Einstein. The spokesman for the group was an Italian engineer who was an enthusiastic supporter of Mussolini, whose recent attempt at raising the purchasing power of the Italian lira dictatorially seemed utopian, to Einstein, in its effect abroad. It was then time for Einstein and Elsa to catch the train to Antwerp, forty minutes away.

There they were met by agents of the Red Star shipping line, who shepherded them to the luxurious Hotel Century—entering through the side in order to thwart the journalists who were lying in wait for them. Einstein and Elsa did not share a bedroom—on account of Elsa's 'fear of snoring,' explained Einstein—and later that evening, they ate their supper in a nearby beer tavern whose ravioli would remain 'unforgettable,' according to Einstein. The patrons of the tavern seemed happily phlegmatic to him, comfortable, and in the fortunate position of being able to eat well. Back in his magnificent hotel room, Einstein took a bath and went to sleep.

After a breakfast on the hotel's thirteenth floor, which afforded a striking view of the city, Einstein and Elsa made their way to the harbor to board the SS *Belgenland*, the liner that would take them and their little entourage to California—by way of New York and the Panama Canal.[8] On this trip, Einstein was accompanied not only by Elsa, but also by Helen Dukas and by Walther Mayer, his mathematical assistant and 'calculator.'

Einstein and Elsa's accommodation on the *Belgenland* was 'princely.' The members of the ship's English crew comported themselves with such elegance and modesty that it made Einstein feel curiously like a country bumpkin. Once settled in his cabin, he was in such a tranquil state of mind that he did not set to work, but read three witty animal stories in a book by Theodor Koch-Grünberg that its author had presented to him.[9]

The *Belgenland*'s first port of call was Southampton. Einstein was on deck to observe the hustle and bustle in the harbor, commenting on how smoothly and calmly the port functioned and that the impressive might of England was manifest. He remarked not only on the gigantic ships and cranes, but also on how the combination of sunshine and a light fog illuminated the silvery sea, and that gulls were soaring gracefully among the countless harbor craft. A number of English reporters came aboard to interview Einstein. He was pleasantly surprised by their restrained good manners: a single 'No' was sufficient for them to desist! The world could learn a lot from the English, mused Einstein; but as for himself, he was not willing to follow their example in everything, and would continue to dress casually, even for the 'holy sacrament of dinner.'[10] Elsa's concerns were of a different nature: she succeeded in having a large bowl provided for their table at mealtimes so that she could dress the salad. The other passengers, apparently, consumed their salad 'naked and dry.'

After another brief stop—in Cherbourg—the *Belgenland* finally reached the open sea and headed west, accompanied by a school of meter-long whitefishes that frolicked alongside her. Einstein settled in his comfortable cabin, and as he resumed his calculations, he chewed on the excellent licorices that Toni Mendel—'bless her'—had kindly provided for his journey.[11]

At the lifeboat drill, the next day (December 5), all passengers had to don their life vests and assemble at their lifeboat stations. The sea grew increasingly turbulent, and the increased motion of the ship drove Helen Dukas to her cabin below deck. Einstein, meanwhile, had a tiff with Elsa over how to respond to the numerous radio-telegrams requesting interviews or lectures that kept arriving over the ship's radio. On a happier note, he made the acquaintance of a fellow passenger, Mr. van Loon, a portly, musical Dutchman, and it did not take long before they were playing violin and piano sonatas together. That evening, Einstein lamented to his diary that one-third of the passage to New York was over already!

The sea continued to be rough for several days, as the ship entered the Gulf Stream in warm and humid weather. Einstein had frequent technical discussions with Mayer, whose systematic work habits earned Einstein's admiration. He also studied the latest developments in quantum mechanics, and was impressed by its spectacular successes. He, nevertheless, regarded it as 'unnatural,' because of his firm belief that a 'good theory' had to grow out of a field theory.

The stormy sea made the ship creak in all her joints and took a heavy toll among the passengers. Einstein observed that older people, he and Elsa included, were less likely to be seasick, just as the ship's doctor had told him. He moaned in his diary about New York drawing ominously close, and then consoled himself with: 'That, too, shall pass.'

On September 9 the ship ran into very strong headwinds and heavy seas. Einstein stepped on the bathroom scale and noted that the minimum and maximum weights it registered were in the ratio of 2 to 3; this allowed him to conclude that the ship's vertical acceleration was approximately 2 m/sec/sec.[12] He also discovered a flaw in the field equations he had been working on and was obliged to discard them. As the sea grew less turbulent, the long-suffering Dukas finally reemerged on deck, but her appearance reminded Einstein of 'a corpse on a brief vacation.'

The next day, his last full day at sea, Einstein composed a message to the Zionist youth of America, which he had agreed to broadcast, and he responded to the countless telegrams that kept the ship's radio operators frantically busy. Einstein's new friend, Mr. van Loon, was of great help in translating Einstein's messages into English.

By late evening the sea had become quite calm, in time for the festive farewell dinner, at which the passengers amused themselves by batting balloons back and forth. Afterward, Einstein and Mayer came on deck to talk and gaze into the starry night sky. They could hear the second-class passengers singing and drinking their final farewell toasts before arriving in the "dry" United States (where the sale of alcoholic beverages had been prohibited for ten years).

FOUR DAYS IN NEW YORK

The tumult that greeted Einstein on his arrival in New York on December 11, 1930, was worse than anything he had imagined in his 'most fantastic expectations.' A horde of reporters and photographers boarded the *Belgenland* while she was off the coast of Long Island, and they pounced on Einstein like a 'pack of hungry wolves.' The reporters asked him extraordinarily stupid questions, to which he responded with 'cheap quips' that were received with gleeful enthusiasm. Once the *Belgenland* had docked at the Hudson River pier at the foot of West 19th Street, the National and Columbia Broadcasting Companies invited Einstein to broadcast a message to the American people. Thanks to a policy cleverly instituted by Elsa, these messages added $1,000 to Einstein's charity chest, which was funded entirely by moneys obtained for autographs and photographs. By noon, Einstein was already feeling half dead. In the afternoon, the chubby German consul, Paul Schwarz, drove Einstein and Elsa around Chinatown and then to one of the East River bridges, where they enjoyed the view of the city's 'wonderful tall buildings' in a light fog. Schwarz then brought them to 'his small apartment, very close to heaven,' which contained little, according to Einstein, other than Schwarz's 'tall, slim English wife who hopped about . . . like a just-caught panther cat.' (The apartment was actually a suite in the Saint Maurice Hotel on 57th Street.)

Einstein spent his first evening in New York in a most agreeable fashion. He and Elsa enjoyed a 'heavenly' dinner at the home of their friend, Leonor Michaelis, but not before Einstein had played several Bach and Brahms sonatas with his host. The two men knew each other well from Berlin. Michaelis was now settled in New York with a professorship at Rockefeller Institute (now Rockefeller University). (The reader may recall that in 1922, in Nagoya, Einstein had had another fortuitous meeting with Michaelis, which also led to a music session.) Following dinner, a group of philosophers arrived to discuss various epistemological issues with Einstein. Then, at long last, he and Elsa returned to their cabins on the *Belgenland*, where guards had been posted to shield them from persistent would-be visitors.

The detailed coverage of Einstein's visit by the *New York Times* makes it possible to fill in Einstein's account of his four days in New York.[13] Even before leaving Antwerp, Einstein had insisted on remaining on the *Belgenland* while in New York and holding no press conferences. He did stay on the ship but was, in the end, prevailed upon to submit to fifteen minutes of 'torture' by journalists and another fifteen minutes with the photographers. Once the time allotted for the photography session on deck was over, Einstein tried to make his escape, but he was pulled back twice by the voracious photographers, who surrounded him. On his third try, he succeeded in getting away and ran for a companionway—only to find the door locked. Finally, he was able to escape into the ship's drawing room. Typical of the barrage of journalists' questions that Einstein parried were these: "Define the fourth dimension in one word." "Define the theory of relativity in one sentence." "What is your view on prohibition? On politics? On religion?" "What are the virtues of your violin?"

It is noteworthy that Elsa, who receives scant mention in Einstein's travel diaries, was treated very sympathetically by the American press. Following the press conference on board the *Belgenland*, the *Times* described her as "serene and unperturbed," as she hovered over her husband. She explained to the reporters that Einstein was afraid of meeting so many people and that he would have preferred not to meet with them. While Einstein seemed phlegmatic to the journalists, who reported that he moved slowly and deliberately, Elsa was said to be "vivacious, quick, deft and very tactful, with the wide-open eyes of a child. Her calm, soft voice was much like her husband's. She is short of stature and dresses in simple, Victorian style." Unlike Einstein, Elsa was said to speak English fluently.[14]

The next morning (December 12), at breakfast, an Indian prostrated himself passionately before Einstein, causing him to flee to his cabin and remain there. Later, the bankers Felix Warburg and Bernard Kahn, both officers of the American Jewish Joint Distribution Committee, called on Einstein. They moaned about the panicky reaction of American Jews to the recent violent demonstrations against Jewish settlers in Palestine (in 1929). They then escorted Einstein to a gigantic Zionist rally, a 'farcical affair' in Einstein's view, which was replete with speeches, films with sound, and a brief, platitudinous response by Einstein himself. Menachem Ussishkin gave a 'fanatical' speech in which his claim, "you belong to us," did not at all sit well with Einstein.[15] This event was followed by a luncheon given by Adolph Ochs, the publisher of the *New York Times*, at which Einstein was obliged to listen to more lofty speeches. He excused himself from responding in kind by portraying himself as being just a naked, indigenous Indian without a gun, armed merely with bow and arrow.

Following a tour of the *Times* newspaper plant, Einstein and Elsa were driven uptown to Riverside Church, where they examined Einstein's likeness carved in stone among the other statues of great scientists and philosophers that decorate the church portal. They returned to the *Belgenland* for dinner, and in the evening, they attended a performance of Bizet's *Carmen* at the Metropolitan Opera, occupying a box together with Dukas and Mayer. When Einstein's presence became known among members of the audience before the overture began, he was applauded enthusiastically. Einstein was unaware that the applause was intended for him, and only after Elsa nudged him did he stand up and wave his handkerchief. After the performance, he met with the celebrated soprano Maria Jeritza in her dressing room, but when a bevy of newspaper photographers tried to ambush him backstage he escaped and made a dash for the waiting limousine. He sat back in the darkest corner of the car and refused to emerge, notwithstanding the urgent pleas of the Met's press representative. He and Elsa were then driven back to the *Belgenland*, their floating refuge.

At a formal reception at City Hall, the next day, Mayor Jimmy Walker presented Einstein with the keys to the city, and Nicholas Butler, the president of Columbia University, welcomed Einstein as a "visiting monarch of the mind" and in a rambling, scholarly speech compared him to Copernicus, Newton, and Kepler. His oration was followed by Walker's 'very witty' speech and then by Einstein's 'brief and wretched' response, in German. He said how grateful he was for the many tributes he received, but he preferred to think that this reception honored not him personally, but all of the scientific research that was being carried out all over the world. The municipal band then struck up the German anthem, "Deutschland über Alles," followed by "Hatikvah" and "The Star-Spangled Banner." When Einstein and Elsa emerged from City Hall, thousands of New Yorkers cheered them outside and they were filmed by a multitude of movie cameras. Their next stop was the Metropolitan Museum of Art, where they visited several galleries and viewed, among other things, Egyptian ceramics and Rembrandt paintings before they were whisked off to a luncheon at the Fifth Avenue residence of Felix Warburg.[16] After the meal, a woman singer presented a lieder recital that gave Einstein particular pleasure—as did most musical interludes. He was then driven to NBC headquarters to read his address to Zionist youth over an international radio syndicate. He urged his listeners not to be discouraged by the current difficulties, but to persevere and to cultivate good relations with the Arab population—for the Palestine they were building must serve *their* real interests, as well; otherwise, the future would be unhappy for Jews and Arabs alike.

Back on board the *Belgenland*, Einstein and Elsa were visited by the Russian cellist Aleksandra Barjanski and his wife, who brought them to the 'harmonious' home of friends, where they spent a much-needed period of quiet and relaxation. Their next public appearance was later that evening, at a gigantic Hanukkah celebration in Madison Square Garden. Einstein and Elsa were wildly cheered by the assembled eighteen thousand Zionists. Einstein gave a short speech, in which he offered words of encouragement in light of the recent Palestinian difficulties. Elsa also spoke briefly and let the audience know how happy she was to be among them for that night's celebration. Both she and Einstein were quickly becoming adept at handling public relations and spouting clichés. It was midnight before they got back to the ship, and Einstein was able to record the day's round of activities in his diary.

On Sunday, December 4, Einstein's last full day in New York, he opened his diary entry with 'Gottseidank!' (Thank God!). The day began with the arrival of several visitors, one of them being Abraham Flexner, an educator who wished to consult with Einstein regarding the organization of a new scientific research institution in Princeton—which would, in due course, become Einstein's academic home. Einstein then called on the great violinist Fritz Kreisler in his hotel, where he also encountered Kreisler's 'Megaera' wearing her nightcap.[17] In the course of their conversation, Einstein learned that several of the popular classical pieces Kreisler claimed to have discovered were, in fact, composed by him. Einstein next called on the celebrated Bengali poet Rabindranath Tagore, with whom he had had several conversations in Berlin the year before.[18] Tagore had recently returned from the Soviet Union, where he had been smooth-talked very effectively, in Einstein's opinion. Tagore had been delighted with what he had been shown—specifically, with the public education system, which he thought would be ideally suited for India.

Later, Einstein and Elsa were the guests of Consul Schwarz at the Hotel Savoy-Plaza, where Einstein was able to continue his chat with Fritz Kreisler until it was time to go to a concert of the New York Philharmonic at the Metropolitan Opera. Einstein and Elsa were again accorded a spontaneous ovation by the audience until Arturo Toscanini stepped on the podium to conduct Beethoven's Sixth Symphony (the *Pastorale*). Einstein was greatly impressed by the measured pace and the clarity of the performance, and he was thrilled to shake Toscanini's hand in the conductor's suite during the intermission. After the concert, he and Elsa were guests at tea in the town house of John D. Rockefeller Jr., who discussed scholarship policies with Einstein. Later that afternoon, Einstein gave one brief address to a delegation of Jewish physicians and another to members of the American Joint Distribution Committee. He then hurried off to the main event of that day, a meeting of the New History Society, at which he delivered a rousing speech in support of militant pacifism. He argued that if just 2 percent of men refused military service there would not be enough jails to hold them, and he urged pacifists to show courage and patience. He declared that as things were, a person was duty-bound to commit crimes in the name of his country, but as far as he was concerned, it was one's duty to liberate mankind from that obligation.

On departure day, December 15, Arthur Fleming drove Einstein and Elsa to the impressive estate of his son-in-law, Lloyd Wilton-Smith, on Long Island. The estate was situated on a wooded hill overlooking a bay not far from the town of Great Neck. Einstein took a liking to his host and family and was astonished by the luxury in which they lived, by the enormous size of their home, and by its covered tennis court. In the evening, he and Elsa returned to the city and to the *Belgenland*, where Einstein was importuned by numerous friends and well-wishers who had come to bid him good-bye. The weather had turned bitterly cold. It was almost midnight before all the farewells and showers of confetti were done. As the ship pulled away from the pier and sailed down the Hudson River, Einstein felt 'a sense of great liberation.'

He had expressed the same sentiment in 1923 when his visit to Palestine came to an end.

AT SEA AGAIN: HAVANA AND PANAMA

After only one day at sea, the weather was noticeably warmer and it was pleasantly humid. Einstein went back to work. Together with Mayer, he studied Dirac's recent papers in which he derived the relativistic wave equation.[19] Einstein also worked on a new exposition of general relativity theory, which he planned to sell to a university press. He discovered that most of the new passengers were old ladies, but with the sea calm and the weather getting warmer day by day, he relished the peace and quiet on board—but 'from Cuba, the Jews are telegraphing already. . . .'

The *Belgenland* arrived in Havana on December 19. Havana, at the time, was in the grip of ongoing civil unrest aimed at Cuba's military dictator, Gerardo Machado. (Two years down the road, Machado would be replaced, with US help, by the equally corrupt Fulgencio Batista.) Einstein commented that although the city was in the midst of a revolution, it was not noticeable. In the wake of the 1929 stock-market crash, the largely foreign-owned Cuban economy had collapsed; world sugar prices had dropped to an all-time low of one-half cent per pound—a devastating blow to the country.

Einstein and Elsa were welcomed ashore by local government officials and intellectuals and by Franz Zitelmann, the local German consul. They were immediately taken on a tour of Havana. In view of the frigid weather he had experienced in New York, Einstein was particularly delighted by the balmy Cuban winter. He admired the city's old Spanish buildings and the absence of 'Americanism' despite heavy American investments. His hosts, on the other hand, left him unimpressed, as they 'dragged him around' from one reception to the next: from the Geographical Society to the Astronomical Society and the Engineering Society, followed by Havana's Hebrew Association—they all seemed the same to him. In the evening, at the Academy of Sciences, Einstein gave a talk in which he laid out his scientific plans for his stay in Pasadena.

What truly appalled Einstein in Cuba was the vast contrast between the luxurious clubs where the receptions took place and the abject poverty of the mostly colored people who lived in windowless shacks. But in spite of severe unemployment and pervasive poverty, Einstein saw many happy faces—thanks to the mild climate and the abundance of bananas. He came to the conclusion that dire poverty exists only where the climate is severe and people are detached from the land.

Einstein had an interview with President Machado the following morning. Later he met with Cuban Zionists, and then with Jewish immigrants who manufactured dresses and undergarments and had brought a children's chorus along to sing for him. A youthful astronomer walked with him to a picturesque fruit market in bright sunshine, before Einstein ended his brief visit to Havana by calling on Consul Zitelmann and meeting his 'clever wife' and their three blonde 'daughters of the Rhine.'

The *Belgenland* departed Havana at 1:00 p.m., but not before Einstein had a most pleasant and utterly unexpected surprise: Jassiko, that 'lovely Japanese child,' now fourteen years of age, came to the ship with her parents to see him off. Einstein had first met little Jassiko and her parents eight years ago when they were fellow passengers on the *Kitano Maru*, en route from Marseille to Japan. Einstein was overjoyed with this unforeseen and happy reunion.

Back at sea, the *Belgenland* followed the Cuban coast before turning south and heading for Panama. The weather turned terrifically hot and steamy. After weathering a series of tropical rainstorms, on December 22 they encountered very heavy seas and powerful winds. Einstein remained in his cabin and continued studying Dirac's recent articles with Mayer. Earlier, he had received word that the message he had broadcast to Zionist youth from New York had met with widespread approval—also from the Arab side, which gave him the greatest satisfaction.

Early the next morning, the *Belgenland* entered the Panama Canal in the midst of great tropical heat. Einstein watched with interest as the ship passed through the three locks, and he followed her progress through the 'fantastic landscape' of basaltic cones copiously clothed in tropical vegetation. As the ship approached the Canal's southern (Pacific) terminus, the basaltic hills on both sides of the Canal grew ever higher, while the air became noticeably drier and appeared to have a bluish hue.

Late in the afternoon, the ship reached the little port of Balboa (today a part of Panama City). A welcoming committee greeted Einstein and Elsa and presented them with a schedule of activities that the German ambassador had prepared for the five hours that their ship would remain in port. First, a swimming demonstration by children—they swam like 'little seals'—then a drive to the top of a hill covered with luxurious vegetation that afforded a glorious panorama. Next, a visit to Panama's president, who turned out to have been a student in Zurich and who happily exchanged youthful reminiscences with Einstein. At sunset, there was a handshaking session on the beach at the German Club, and then a cheerful Berliner—a Frau Kuhn— drove the Einsteins to the ruins of the original site of Panama City, sacked by the privateer Henry Morgan in 1670.[20] Einstein and Elsa then ended up in a 'fantastic garden' where they dined *al fresco* to the sounds of a melancholy jazz band. The president of the Chamber of Commerce presented Einstein with an exceedingly precious panama hat, and by the time the entire party returned to the *Belgenland*, everyone was in animated high spirits. In his diary, Einstein professed that those few hours ashore had indeed been 'magical.'

In milder weather and under a starlit sky, the *Belgenland* put out to sea once more. For Christmas Eve, a concert was organized on board in which Einstein played the violin. Two days later, while the ship steamed inexorably northward along the Mexican coast, Einstein lamented that his sea voyage would soon come to an end; but quickly he consoled himself with the thought that the solitude he enjoyed at home, in Caputh, was even lovelier!

The heat was again so intense that Einstein was reminded of purgatorial fire. As if that were not bad enough, the passengers were getting increasingly brazen with their endless requests for photographs—although their photographing did add many dollars to Elsa's charity chest. Einstein, moreover, felt that he had made an ass of himself by agreeing to broadcast a Christmas message to Americans from the ship. Clearly disturbed by all the fuss, he moans: 'how will this end?'

A few days later, he enjoyed a spectacular, flaming sunset—something he never tired of—from the deck while masses of small hogfishes (*Bodianus pulchellus*) danced on the water alongside the ship. He paid a visit to the captain on the bridge, a 'nice chap with a wide-awake roguish grin,' and put up with the inevitable photographing that his visit precipitated. A concert that the passengers staged that evening left Einstein grumbling about their childish humor and artistic mediocrity.

By December 29, the weather had turned decidedly brisk, and it was necessary to wear an overcoat while on deck. Together with 'Mayerchen,' Einstein returned to the study of Dirac's recent profound contributions to quantum mechanics—'a truly ingenious system,' according to Einstein. San Diego was only two days away, prompting Einstein to confess to his diary that he was a little afraid to set foot on this land of unlimited possibilities! After a whole month spent on the *Belgenland*, the time had come for Einstein to bid farewell to his fellow passengers, but only his farewell to the ship's musicians touched him deeply.

PASADENA

The *Belgenland* made land in San Diego early in the morning of December 31. Elsa woke her husband with the news that a throng of journalists was waiting for him on deck. Following the press conference on board, at which he fielded the usual questions, Einstein was welcomed by the mayor, and he was then serenaded by trumpeters and a chorus of boys and girls. Afterward, all the girls came aboard, each carrying a bunch of red flowers, which they presented to Einstein, who soon held so many flowers that he could barely get his arms around them. Finally he fled to his cabin. After a while, he and Elsa made their way through the huge, waiting crowd and were taken on a sightseeing tour of San Diego. Einstein was pleased to discover that the city had retained some of its old Spanish charm, but what truly astonished him was that in this city, there was one automobile for every two citizens!

At the formal welcoming ceremony in the City Garden, the rabbis of both of the city's Jewish congregations delivered welcoming speeches. When the organist played, in place of the German national anthem, "Die Wacht am Rhein" ("The Watch on the Rhine"), Einstein and Elsa were bemused because it was the theme song of militant German nationalists. (Five years earlier, he had encountered the same musical gaffe at a reception in Montevideo.) Following a hotel luncheon, Einstein's Caltech host, Arthur Fleming, drove the pair to Pasadena. During the three-hour trip, Einstein was thrilled whenever the coastal road allowed views of the dunelike hills that ranged far into the distance.

Elsa and Einstein spent their first night ashore in Fleming's 'wonderful Swiss chalet' with its magnificent, steeply sloped garden. The next day was New Year's Day, the day on which the annual Tournament of Roses parade takes place in Pasadena. The Einsteins were invited to join Fleming to watch the parade from a well-situated bank downtown, where they arrived with a police motorcycle escort. Seated at a large window looking out on the parade route, Einstein was duly impressed by the multitude of marching bands and by the many fantastic floats decorated with thousands of flowers. According to one *Los Angeles Times* reporter, it was 'one of the most delightful days [Einstein] had experienced.'

The next morning, Einstein visited Caltech's physics department and later strolled in Fleming's garden. Work in earnest began the next day. Although Einstein was initially skeptical of Tolman's cosmological model, presumably the "pulsating universe," Tolman was able to convince him of its correctness.[21] That afternoon, he and Elsa moved into their own Pasadena residence, a 'gingerbread house with a shingle roof,' which would be their home for the next two months. Their move was closely observed and photographed; Pasadena's 'rag sheet' informed its readers the very next day that Einstein was at work in his study within an hour following the move. Elsa, meanwhile, received a visit from Mrs. Barbara Seibert, a girlhood friend in Germany, and the two women went shopping together.

Einstein and Elsa spent the evening of January 6 at the home of Robert Millikan, who had achieved fame for measuring the charge of the electron, and was venerated like a god in Pasadena, according to Einstein. The next day, Einstein, Millikan, Albert Michelson, and Walter Adams were filmed together for a newsreel.[22]

Einstein described Pasadena as a huge garden with a rectangular grid of streets. The streets were lined with villas, each surrounded by a garden containing palm trees, small-leaved oaks, and pepper trees—which Einstein held responsible for all the dishes being over-peppered in Pasadena. The town was surrounded by round-topped hills covered with tropical vegetation. Cutting through the landscape were straight roads, along which the inhabitants traveled in long columns of cars. The important role played by cars astonished Einstein, as did the large number of 'prehistoric rattletraps' he saw. He came across one such dilapidated jalopy for sale for only $25! He noted that he rarely saw pedestrians in Pasadena, but everyone he did encounter on the street knew who he was and smiled at him. He discovered that the local stores operated in a wonderfully ingenious manner, unknown in Europe: you took a basket when you entered, and when you left, you paid for whatever you, yourself, had placed into it. Einstein was also greatly impressed by the ingenious packaging he encountered, and specifically, by the egg cartons.

At the theoretical physics colloquia, which he attended regularly, Einstein took his place among the students and professors and smoked his pipe. In his colloquium, he used a wooden box as a stage prop to illustrate a thought experiment; this was, most likely, another attempt to demonstrate a fundamental inconsistency in the Heisenberg uncertainty principle. He also took part in the physics and astronomy colloquia; in one of them, he heard Walter Adams talk about his latest attempts to measure the spectrum of Sirius B, the supermassive companion of Sirius A—"the Dog Star" in the constellation Canis Major. That experiment was of particular interest to Einstein because it had the potential of demonstrating the gravitational redshift that was predicted by general relativity.[23] At another colloquium, Charles St. John's account of the rotation and circulation of the sun inspired Einstein to propose a thermodynamic explanation for St. John's observations.[24] Einstein was delighted by the congenial (*sympatische*) atmosphere that permeated all of the academic gatherings. He clearly thrived in Caltech's invigorating and informal environment, so different from the far more rigid atmosphere at the Prussian Academy.

Hollywood is only a short distance from Pasadena, so it was inevitable that the celebrities from the realms of film and science would find each other. Einstein had his first encounter with Hollywood on January 8, when he met with the movie producer Carl Lämmle, a Swabian Jew like himself. Einstein described him as a clever little hunchback who 'lets the film stars dance and is a master of the humbug.'[25] Lämmle was a celebrated film czar, the founder of Universal Studios, which had recently released the antiwar film *All Quiet on the Western Front*. The film had been banned in Germany, and Einstein recognized that the ban represented a major diplomatic defeat that revealed the weakness of the Weimar government and the growing influence of the Nazis. Lämmle invited Einstein, Elsa, and Helen Dukas to a special screening of the film. During the showing, the lights came on and Mary Pickford, the celebrated silent film star, walked down the aisle to introduce herself to Einstein, who politely shook her hand. Afterwards, Dukas overheard him asking Elsa, 'Who is Mary Pickford?'[26]

Charles Chaplin, on the other hand, was a film star Einstein was familiar with and admired. Hearing of Einstein's interest, Chaplin invited him and Elsa to dine with him and his then-partner, Claudette Colbert. Chauffeured to dinner in Chaplin's automobile, the Einsteins spent what must have been a fascinating evening in the company of Chaplin, Colbert, Lämmle, and the publisher William Randolph Hearst and his companion, the actress Marion Davies (whose portrayal in the film *Citizen Kane*, in the dippy character of "Susan Alexander," seriously maligns her). In the course of the dinner, Chaplin invited Einstein and Elsa to join him at the world premiere of his latest film, *City Lights*, which was to open later that month. They accepted his invitation, but at that glittering affair, Einstein was baffled by the brouhaha and the crowds that greeted the two celebrities when they arrived for the opening. He asked Chaplin what it all meant. "Nothing," was Chaplin's reply.

On January 9, Einstein visited Upton Sinclair, a prolific and popular novelist and social critic of his times, who had just won the Nobel Prize for literature—the first awarded to an American. Although he is somewhat neglected today, he was a severe and influential critic of American society in his time. Einstein regarded him as a magnificent idealist, noting that despite that, he had retained his cheerful personality. Sinclair was favorably disposed toward the Soviet Union because of its efforts to awaken and educate the masses. He invited Einstein to view some film footage that the celebrated Soviet film director Sergei Eisenstein had recently shot in Mexico. This footage was all that remained of an ill-fated film project that had mostly been financed by Sinclair and his wife. Eisenstein had shot some 200,000 feet of film in his several months–long stay in Mexico, during which he had mixed socially with the artists Frida Kahlo and Diego Rivera. Eventually, the project collapsed amid bitter controversies involving Eisenstein, the Sinclairs, and Eisenstein's actual employer, Joseph Stalin.[27]

Millikan and others in Pasadena's academic establishment kept Einstein busy with a string of testimonial gatherings and fundraisers that usually took place in Caltech's Athenaeum. That was also the venue where Edna Michelson, the wife of Albert Michelson, gave a luncheon in honor of Elsa, whose magnificent supporting role to Einstein was explicitly recognized on that occasion. Even if she features little in the travel diaries, her vivacious presence and her care for "Albertle" is manifest and can be discerned in photographs.

On January 15, an extraordinary dinner in honor of Einstein and Elsa took place in the Athenaeum, at which seven distinguished scientists gave popular presentations of the recent dramatic developments in physics and astronomy.[28] Their talks, together with Einstein's response, were broadcast coast-to-coast. Speaking in German, Einstein thanked Millikan, his host, and said that he felt fortunate to be breaking bread with his fellow scientists in such a joyous mood. He then pointed out that he had been only three feet tall when Michelson performed his 'marvelous experimental work,' which kindled the ideas of Lorentz and Fitzgerald and paved the way for his theory of relativity. Michelson thanked Einstein for his gracious remarks and commented that he had no conception in 1887 that his experiment would have such far-reaching consequences. The other speakers at this star-studded event discussed the photoelectric effect, the expanding universe, and the latest experiments designed to put the predictions of relativity theory to the test.

It is a reflection of Einstein's busy academic and social schedule that he omitted making daily diary entries after January 9 and did not resume them until January 22. By then he had come to the conclusion that the Pasadenians (*die Pasadener*) were as fond of pomp and ceremony as he had found the Spaniards to be. He noted that this was particularly true of Greta Millikan, who reminded Einstein of an evangelical pastor. She often appeared at Einstein's house to present him with a social program that she had prepared for him, including his worst assignment, a reception for the financial benefactors of Caltech. On that occasion he had been obliged to shake 350 hands and to listen to numerous unctuous speeches, including one by himself.

But scientifically, Einstein continued to savor the stimulating environment at Caltech, where he was surrounded by scientists who provided the experimental foundation for his theoretical work. He was anxious to visit the Mount Wilson Observatory, but first he had to submit to an examination by a physician to make sure that the observatory's elevation (5,700 feet) did not pose a health risk for him. Edwin Hubble then drove Einstein and Elsa up the winding road to the observatory, where they were shown the gigantic 100-inch telescope. Einstein was also able to examine the photographic glass plates on which the spectra of distant galaxies and stars were recorded. He also talked with Charles St. John, who showed him some recently discovered solar phenomena; for example, the observation that the magnetic fields of sunspots rotated in opposite directions following each revolution of the sun.[29] These findings intrigued Einstein sufficiently to search for their theoretical explanations.

Before Einstein came to Caltech, it was widely believed that he would consult with a few theoreticians and would otherwise be a virtual recluse. Far from that expectation, he spent his days visiting colleagues and their laboratories and displayed an informed interest in technical details—for example, in the construction of thermocouples that were so sensitive that they could measure the heat emitted by distant stars.[30] His interest in such practical matters was not surprising for anyone aware of Einstein's long involvement with inventions and patents in general and with the gyrocompass in particular. At the same time, he had productive interactions with theoreticians, particularly with Richard Tolman, whom Einstein regarded very highly. Tolman, along with Hubble, was able to convince Einstein to abandon the original assumption of relativity theory that the universe was static and uniform. Indeed, Einstein complained in his diary that his colleagues at Caltech were so interesting and so gracious that he could hardly find time for his own work.

It was not all work, however. A local instrument dealer lent Einstein a marvelous Guarneri violin, which he made use of as often as he could. He found an excellent chamber-music partner in the once-celebrated violinist Lili Petschnikoff (of whom more in chapter 7) and he often played in her home. He and Elsa also attended a concert of the Los Angeles Philharmonic, and after hearing a performance of the Fourth Symphony of Brahms, Einstein had high praise for the conductor, Artur Rodzinski.

Going even farther afield, Einstein and Elsa spent several relaxing days in Palm Springs, as guests of Samuel Untermeyer, a wealthy New York lawyer and generous supporter of Democrat and Zionist causes. After a 'wonderful' drive through the California desert, they arrived at Untermeyer's sumptuous mansion (January 25).[31] They spent a very pleasant evening with their host, whom Einstein had known for years, although in Einstein's opinion, he had become much too pessimistic over the years. In the morning, Einstein climbed a steep hill behind the Untermeyer house to view the sunrise: as the sun rose over the mountainous desert, the leaden-gray rock ledges of the San Jacinto Mountains were turned into pure gold. And before long, everything was bathed in glorious, warm sunshine, so dear to Einstein's heart. Later, Untermeyer drove the couple to a cactus farm, where the fantastic shapes of the cacti intrigued Einstein, but the ubiquitous photographers less so. From there they drove to Palm Canyon and climbed down into the deep gorge where a small stream made it possible for palms to flourish in the middle of a barren desert. Many other families had also come to the canyon for a Sunday outing, and even in the middle of the desert they all knew who Einstein was. They incessantly wanted to shake his hand and be photographed with him—all, except for a little girl with a 'malevolent frown' who refused to be photographed with him. 'With God's help, she'll remain that way!' commented Einstein.

The next day, Untermeyer drove his guests on an excursion through the desert to a wonderful plantation where dates, grapefruit, and vegetables were grown; and where, at day's end, they were treated to another magnificent desert sunset.

*   *   *

Regrettably, here Einstein's travel diary comes to an abrupt halt. He and Elsa remained in their little house in Pasadena until the end of February 1931, when they traveled to New York, arriving via the Manhattan Limited at Pennsylvania Station in the early morning of March 4. Their ship, the SS *Deutschland*, was to depart at midnight.[32] In that brief intervening time span, Einstein was able to lend support to his two favorite causes: pacifism and Zionism.

After Einstein and Elsa had breakfasted on board the *Deutschland*, Einstein gave a fervent speech to a delegation of the War Resisters' League, which included the Socialist leader Norman Thomas. Thomas asked Einstein whether the many bread lines in the United States had not appalled him; to which Einstein replied that today's social and economic problems were much more difficult than the problem of peace. He also succumbed to Weizmann's entreaties to appear at a fundraiser in the Astor Hotel in the evening. Twelve hundred people had paid $100 each to see and hear the great man—a considerable sum in the midst of the Depression. In his address, Einstein again urged Jews to cooperate with the Arabs in Palestine and to consider the example of Switzerland, also a country composed of several distinct national groups.

Just before the *Deutschland* sailed, a reporter asked Einstein how he reacted to the acclaim he had received from millions of people who understood nothing of his work; he replied that he was so absorbed in solving the riddle of the universe that he could not be expected to try to solve the riddle of humanity as well.[33]

After a rough Atlantic crossing, the *Deutschland* arrived in Cuxhaven, Germany, on March 14. When Einstein saw the masses of reporters and batteries of cameras that waited for him ashore, he asked the captain to let him and Elsa remain on board rather than disembark with the other passengers. They were the only passengers on board as the ship sailed up the Elbe River to Hamburg, whence they traveled directly to Berlin.[34]

Einstein would spend two more winter sessions at Caltech. His stay of 1931–1932 is covered in chapter 7, and that of 1932–1933, in chapter 8. Einstein thought both these subsequent visits were more enjoyable and productive than the first, in large part because public interest in Einstein's person was, mercifully, much more subdued.

But before embarking on his next American journey, Einstein traveled to another academic refuge of his, Oxford.

6.

Berlin and Oxford (1931)

In the spring of 1931, Einstein was invited to present the prestigious
Rhodes Memorial Lectures in Oxford and to spend a month as a visiting
scholar at the university. The invitation had been instigated by the physicist Frederick Lindemann, who had first met Einstein and become his unwa-
vering admirer during the 1911 Solvay Conference, when the cream of theoretical physicists assembled in Brussels. The scientific paths of the two men
had, however, crossed earlier: in 1907 Lindemann, who was brought up in
Britain and had gone to Germany to study physics, was working in Nernst's
laboratory in Berlin. There he made the specific heat measurements of
metals at very low temperatures that confirmed the predictions of Einstein's
quantum theory of metals.[1] After he returned to England, he made important
contributions to aeronautical science during the First World War, when he
was the director of a Royal Flying Corps facility. After the war, he was
appointed professor of experimental philosophy at Oxford.

By 1931, Einstein was becoming increasingly aware of the growing
menace that Hitler represented, and he had begun to explore academic positions at Caltech and in Princeton—in case he decided to leave Berlin. Lindemann's proposal offered Einstein an opportunity to renew contacts with
British scientists and to communicate to them his political apprehensions. He
accepted the invitation with pleasure.

Contrary to custom, on this occasion Einstein began keeping a travel
diary twelve days *before* his departure. The entries he made between April 8
and 20, therefore, offer a rare glimpse into his daily life in Berlin.

AT HOME IN BERLIN

Einstein's first entry tells us that April 8, 1931, was a lovely spring day in
Berlin. The writer was sitting in his small attic study, the so-called *Turmz-
immer* (tower room), one flight above the apartment he shared with Elsa on

117

the Haberlandstrasse; he was savoring the tranquility he enjoyed there. The study had been specially constructed for him and was furnished very simply, containing just a bookcase, a small desk, and a chair. On its walls were pictures of Einstein's most admired predecessors: Isaac Newton, Michael Faraday, and James Clerk Maxwell.

Earlier that morning, Einstein had finished an essay about Maxwell's conception of physical reality, on the occasion of the great physicist's one-hundredth birthday.[2] He then began working on a manuscript that presented a systematic catalog of possible field equations for a 'Distant Parallelism' unified field theory. In the diary, he reminded himself that the following day, he would travel to the astronomical observatory in Potsdam to help calm the latest squabble between the observatory's director, Hans Ludendorff, and the physicist Erwin Freundlich. He would be joined in this by Max von Laue and Erwin Schrödinger. Einstein referred to the trio as the three 'peace doves,' and their mission throws light on Einstein's academic activities in Berlin.

Hans Ludendorff, the director of the Potsdam observatory, was the brother of the infamous General Erich Ludendorff, who, along with Hindenburg, had held the highest military command in the First World War. General Ludendorff was a dedicated German nationalist who had taken part in Hitler's failed Beer Hall Putsch in 1923, an action he came to regret profoundly ten years later.[3] Hans Ludendorff clashed frequently with Freundlich, who was the director of the Einstein Institute—the solar observatory that was housed in the futuristic "Einstein Tower." The instrumentation of the solar observatory had been specially designed by Freundlich for experiments to test the validity of general relativity and specifically to measure the gravitational redshift of sunlight. Because of the tremendous turbulence at the sun's surface, that experiment turned out to be far more difficult to implement than was anticipated, and the redshifted spectrum of Sirius B now offered a more promising approach (see chapter 5). Although the Einstein Institute was nominally under Ludendorff's overall control, it enjoyed a measure of autonomy because part of its funding came from the private Einstein Foundation.[4] This ambiguous administrative structure led to unending disputes between Ludendorff and Freundlich—disputes that were exacerbated by the autocratic style of the former, and the high-handed manner of the latter. The scientific outlook of the two men differed profoundly as well, for Ludendorff was an observational astronomer with little interest in theory, while Freundlich had studied theoretical physics in Göttingen.

All three 'peace doves' were trustees of the Einstein Foundation, and von Laue and Schrödinger were among Einstein's most congenial colleagues at the academy and the university. Laue, a former student of the revered Planck, and a close friend of Einstein's, defended relativity theory even after the Nazi takeover, and he was the only academician to object to Einstein's expulsion from the academy in 1933.[5] After the Second World War, he was the only German scientist with whom Einstein was willing to correspond.

Schrödinger, the third peace dove, was an Austrian physicist who had been enticed to Berlin from Zurich in much the same manner as Einstein was a decade earlier. His own annus mirabilis occurred in 1926, when he published a group of six papers that laid the foundations of wave (or quantum) mechanics. Soon afterward he was offered the chair in theoretical physics at Berlin University and was elected to the Prussian Academy. After he arrived in Berlin, he and Einstein became friends. They shared an aversion to the stiff formality of their colleagues at the academy, preferring to dress informally, and neither of them held monogamy in high esteem. Schrödinger was a frequent guest at Einstein's retreat in Caputh, where they sailed and took long walks together. Unlike Einstein, Schrödinger took no active part in German politics, and although he was not Jewish, he gladly exchanged his professorship in Berlin for a fellowship at Oxford University when Hitler came to power in Germany in 1933.[6]

On this occasion, the three professors succeeded in restoring peace. (The uneasy relations between Ludendorff and Freundlich sputtered on, until the tension was finally resolved under the Nazi government: since Freundlich had a Jewish grandmother, he was forced to resign his position and emigrated. Ludendorff, meanwhile, became a ready ally of the Nazis and gained control of Freundlich's Einstein Institute.)

His mission in Potsdam completed, Einstein spent a very enjoyable evening with the economist and sociologist Franz Oppenheimer. Einstein admired Oppenheimer's brilliance and wit, and he was particularly taken by the superb jokes Oppenheimer told that evening—alas, he failed to record them in his diary.

In the days remaining before he left for England, Einstein worked each day on his current approach to a unified field theory, either alone or with his 'calculator,' Mayer. But he also performed his academic duties and engaged in extensive social, political, and musical activities. Thus, on April 11, he called on Wolfgang Windelband, a historian who represented the Prussian

Ministry of Education, to press his request for an academic appointment for Mayer. Einstein had come to depend on Mayer as a collaborator and had even declined an offer from Caltech because it did not also provide for Mayer. Later that day, he received a visit from a nephew of Oppenheimer, a 'sympathetic fine boy,' a gardener who wanted to emigrate to California. Oppenheimer's nephew had come to seek advice from one who had been there.

The following morning, Einstein played chamber music with friends: Ms. Herrmann played first violin, Erna Schulz viola, and Ewel Stegmann cello. With Einstein playing second violin, they read through a quartet by Brahms and a Mozart divertimento. It is worth noting that Brahms's string quartets are technically very demanding and that Einstein's three companions were experienced, professional musicians. Erna Schulz was the violist of the prestigious Wietrowetz Quartet and would shortly play with Einstein again, in Oxford. It is a gauge of Einstein's priorities that in the midst of his many commitments, he took part in at least six chamber-music sessions during the last two weeks he spent in Berlin. Mozart, Schubert, and Brahms were his preferred composers.

In the afternoon, Einstein and Freundlich visited Erich Mendelsohn, the architect of the Einstein Tower, in what Einstein described as Mendelsohn's 'tasteful, exceedingly American home.' Einstein's appetite for chamber music was not satisfied, and, after dinner, he met with two companions to play Haydn piano trios. Afterward he climbed up to his attic study and began an article on the cosmological problem—what general relativity tells about the universe.

Einstein's voluminous correspondence took up much of the next day. Among the letters he dictated to his secretary, Helen Dukas, was a petition to President Masaryk of Czechoslovakia, on behalf of a young man who refused military service. He then completed the cosmology manuscript he had begun two days earlier and submitted it to the academy. In the article, Einstein accepted the dynamic (rather than static) model of the universe as the correct solution of his gravitational field equation and abandoned the cosmological constant as unnecessary.[7]

The chain of events that brought Einstein to this step is briefly as follows. In 1917, when he first applied the gravitational field equation to cosmology, most astronomers envisioned the universe as a static sphere encompassing our galaxy, the Milky Way. It was therefore natural for Einstein to include in the equation a term that had the effect of preventing the eventual gravitational collapse of the universe, a term known as the cosmological constant. It was not until the 1920s that other galaxies were identified and only in 1929 that Edwin Hubble published his finding that all galaxies were receding from each other, at speeds proportional to their separation (Hubble's Law). Indeed, only three months earlier, during Einstein's visit to the Mount Wilson observatory, Hubble had shown him the spectroscopic evidence for the expanding universe (see chapter 5). *Without* the cosmological constant, the solution of the field equation represented an expanding universe originating from a big bang—a solution that the Russian physicist Aleksandr Friedmann had first proposed in 1922 and which is essentially the accepted view today.

Einstein is said to have remarked at that time that introducing the cosmological constant had been his greatest blunder. But modern astrophysical data have shown that the most distant galaxies recede at speeds *greater* than predicted by Hubble's Law, an observation that has been explained by a hypothetical "dark energy" permeating all of space. The relationship between the cosmological constant and "dark energy," which accelerates the expansion of the universe, is far from settled and continues to be debated today.

On the day he submitted his cosmological paper (April 16), Einstein also attended a meeting of the academy and listened to a 'stupid talk' by a philologist, a 'typical bagatelle.' The speaker had investigated the source of the funds that Emperor Augustus used for certain public donations. Einstein asked in his diary whether this was really what the state paid bookworms to do.

Einstein spent the evening at a 'splendid performance' of Lessing's play, *Minna von Barnhelm*. He particularly enjoyed the performance of the actress Käthe Dorsch; he found it interesting that the play showed that even Lessing was not free from certain class prejudices.[8]

During the next few days Einstein completed the manuscript that presented a catalog of potential field equations for a unified field theory and asked Mayer to check and proofread it. In the evenings, he either played chamber music or pursued the busy social life he and Elsa engaged in. They attended several dinner parties, at which their fellow guests included family members, academics, politicians, businessmen, artists, and, on one occasion, Berlin's police commissioner.

On April 19, Estella Katzenellenbogen, the owner of several flower shops in Berlin and a frequent companion of Einstein's, invited him to the theater. They saw *The Blue Boll*, an expressionist drama by the sculptor and playwright Ernst Barlach, best known for his powerful antiwar stance in the aftermath of World War I.[9] Two days later, Einstein mentioned spending 'noon and afternoon' with his longtime friend Toni Mendel. (Elsa acquiesced, perhaps not happily, to the close friendship between Einstein and Toni, who was the widowed mother of Hertha Mendel; in any case, the Einstein and Mendel families remained on good terms.[10]) Together they read and discussed Freud's recently published essay *Das Unbehagen in der Kultur* (*Civilization and its Discontents*), which deals with the tensions between the individual and society and which had caused a considerable stir. Einstein admired Freud's writing style but remained skeptical of his psychoanalytical theories. Freud and Einstein, nevertheless, maintained a warm and often lighthearted correspondence for many years.[11]

On April 23, Einstein listened to a lecture at the academy by Fritz Haber on catalysis in aqueous solutions, and then he attended a 'memorable' faculty meeting in which Planck showed great skill in nominating Laue for the Solvay professorship. Einstein also visited the home of Berthold Israel, a wealthy Berlin merchant, where he heard a man and a woman give a 'wonderful' performance of Negro spirituals.[12] He then called on the aging writer Alexander Moszkowski, who had become blind and lame but was mentally still alert. (Moszkowski had first met Einstein in 1916, well before he became world famous, and had published the earliest book about Einstein, in 1920. The book was based on a series of conversations, and its publication was stressful for both the author and his subject. Einstein disavowed the book, but in the end, Moszkowski and his wife, Bertha, remained close friends of Einstein and Elsa's.[13])

A few days earlier, Einstein had lunch with Henry Goldman, the New York banker, who was happy to hear that Einstein had not accepted Caltech's offer; he thought that Einstein's liberal and pacifist views would have made an eventual conflict with the conservative Millikan inevitable. In the afternoon, Einstein went shopping for a piano for his sister, Maja—the two had always been very close. He settled on a Blüthner, an instrument highly prized for its fine sound.[14] In the evening, Einstein was scheduled to address a meeting of the Human Rights League in support of the mathematician Emil Gumbel, who had been put in serious danger by publishing an anti-Nazi pamphlet,[15] but Einstein never had the opportunity to read the speech he had prepared.

Einstein began the morning of April 29 writing numerous letters with the help of Dukas, among them one informing a newspaper that his pacifist ideas had received support from American clergymen. Later that day, he began his journey to Oxford by boarding the evening train to Hamburg, where he would embark for England. Max Warburg, the Hamburg banker, happened to be on the same train, and the two men used the opportunity to discuss various issues related to Palestine and, specifically, to the Technion in Haifa. When they arrived, Einstein was met by a Hapag agent who escorted him to the nearby Hotel Reichshof, located conveniently close to the ship harbor on the Elbe.

Einstein embarked the next morning on the SS *Albert Ballin*, a luxuriously appointed ocean liner equipped with stabilizers, tennis courts, and even an outdoor bowling alley. It was the first passenger liner built in Germany after World War I. The name of the ship must have given Einstein a twinge, for Albert Ballin, Hapag's managing director until 1918, had been one of Kaiser Wilhelm's most devoted *Kaiserjuden*—that group of wealthy, influential Jews whose advice the kaiser often sought. Their counsel, however, often went unheeded, as when they urged him to curb official anti-Semitism or tried to keep him from rushing headlong into war.[16]

Ensconced in a magnificently appointed cabin, Einstein was delighted to be at sea again in lovely weather, and to his diary he bemoaned the brevity of this journey. The unified field theory was very much on his mind; he felt sure there must be field equations that were consistent with the earlier gravitational equation.

Einstein resolved to work toward uniting the clergies of all nations with the aim of legalizing the refusal of military service.

OXFORD: COLLEGE LIFE

The *Albert Ballin* arrived in Southampton on May 1. Lindemann came aboard and brought Einstein ashore in a small motor launch, and the two men then drove to Oxford in Lindemann's chauffeured Rolls Royce. Along the way, Lindemann, who spoke German fluently, told Einstein about the student stipends that England was providing for young people who had talent but no money to allow them to study at a university. They made a detour in Winchester to let Einstein see the town's lovely Gothic cathedral and to pay a brief visit to Winchester College, founded in 1382 and probably Britain's oldest extant boys' school.

At the college, the pair was given a short tour by one of its pupils, John Griffith, who many years later wrote an account of the unexpected visit by "the immaculately attired Lindemann, attended by a short figure, dressed in a Middle-European-style cape with frizzy hair escaping from a kind of skull cap." Winchester College is housed in ancient buildings whose walls carry many commemorative marbles, and these evidently intrigued Einstein. In a room adorned with such marble plaques that was used as a changing room, with sports attire hanging on pegs along the wall, Einstein reflected for a while and then commented to Lindemann: "Ah, I understand, the spirit of the departed passes into the trousers of the living." (*Ach! Ich verstehe: der Geist der Gestorbenen geht in die Beinkleider der Lebenden hinüber.*)[17]

After resuming their journey, the two arrived in time for dinner at Lindemann's college, Christ Church, Einstein's residence in Oxford. Einstein dined in the basilica-like hall, along with some five hundred dons and students, all in tuxedos and academic gowns, a 'bizarre and boring affair' for which he had been obliged to don his despised tuxedo. The meal, which was, 'naturally,' served only by men (women were allowed at the college only on rare occasions), gave Einstein an intimation of how ghastly life without womenfolk would be.

The suite of rooms Einstein occupied at Christ Church normally belonged to a philologist who was currently away in India. The suite reminded Einstein of a small fortress. He enjoyed his quiet, monastic existence there, but not the freezing cold in his rooms. He even came to enjoy the daily 'hallowed evening meal' at which he was ceremoniously introduced to the other members of his 'band of tuxedo-clad brethren . . . [who] were taciturn, but kindly, with delicate little jokes on the tips of their tongue.'

On May 4, Einstein was invited to lunch at the lovely country home of Gilbert Murray, the celebrated classics scholar and politically active humanist. Einstein discussed the deplorable international situation with him and was surprised by the erudition of the other luncheon guests. He was particularly impressed by the women, who, he noted, took a far greater interest in public affairs than women back home.

The loveliness of Oxford's environs did not escape Einstein's notice. He went on many random walks and was delighted by the beauty of the countryside and by English architecture: nothing pompous; everything well-considered, tasteful, and conspicuously in accord with tradition. He also deemed it characteristically English that it was customary in college upon encountering someone in the courtyard or a hallway not to exchange greetings but to adopt, instead, a certain measured attitude—a custom that Einstein found powerfully evocative.

On May 5, Einstein dined at Trinity College and afterward was part of a small group that heard the astrophysicist E. A. (Arthur) Milne explain his theoretical model of the interior structure of stars. (At the time, the energy source of stars—nuclear fusion—was not known.) Milne—a very clever man, according to Einstein—also discussed novae, which had only recently been discovered. He said that their explosive metamorphosis into white dwarfs was caused by the nuclei losing all of their electrons, and the stellar gas becoming "degenerate"; as a result, the stars became much smaller and denser. The thermodynamics of stars was evidently a new interest for Einstein, as was Milne's new 'kinematic relativity,' a theory he proposed as an alternative to general relativity to explain the expansion of the universe.[18] During the next few days, Milne and Einstein met several more times in Einstein's college rooms, sometimes with Lindemann also present; they talked about the cosmological problem and the matter density of the universe in view of the latest astronomical data for the Milky Way.

Einstein relished his quiet existence at college, as well as his walks in the cool, occasionally sunny weather. In the course of one morning's walk, he discovered a particularly interesting set of field equations for representing gravity—but there is no further mention of them in his diary. One day he was invited to a house concert at the home of local music lovers, the Pearces, where he heard Adolf Busch and Rudolf Serkin, the most celebrated violin-piano duo of that time, perform very beautifully some Beethoven sonatas and a rondo by Schubert.[19] After the concert, Einstein was escorted home by Erna Schulz, the violist with whom he had played in Berlin two weeks before.

Einstein gave his first Rhodes Memorial Lecture on May 9. He discussed relativity theory from the point of view of logic—a strenuous exercise, commented Einstein in his diary, since he spoke without notes. In the afternoon, Lindemann drove him to the country house of his father, and Einstein was delighted to make the acquaintance of this 'lively spirit,' despite his eighty-five years. The elder Lindemann had been a businessman in czarist Russia and told Einstein of the all-pervasive corruption he had encountered there: after he had been paid for nine of the ten ships that his firm had delivered to Russia, he asked the government official in charge of the sale why he had not received payment for the tenth ship. The official replied, "And what about me? See to it that you get home as quickly as possible!" Count Sergei Witte, then Imperial Russia's prime minister, had also been exceedingly corrupt, exacting huge bribes in return for mining concessions, according to the elder Lindemann. Einstein discovered with pleasure that his host was an amateur astronomer and had, together with his son, designed and built a number of very ingenious scientific instruments. Their electrometer and galvanometer— instruments for measuring electric charge and current—were especially 'pretty,' in Einstein's view. He was, furthermore, delighted with his host's excellent cuisine, his garden, his little dog, and the view of the Thames valley from his house—all in all, he pronounced it a most successful visit.

MUSICAL OXFORD

Although Einstein must have been aware of Oxford's lively musical scene, he had not brought his violin with him. He was, nevertheless, able to enjoy several evenings playing chamber music on a borrowed violin, several of them at Gunfield, the home of the Deneke family. The Deneke sisters, Margaret and Helena, early benefactors of Lady Margaret Hall, the first women's college of Oxford University, played an important role in Oxford's musical life. Their mother had been a close friend of the celebrated pianist Clara Schumann, and both her daughters had studied piano with Clara's daughter Eugenie; Margaret had gone on to become a concert pianist. It was in the Gunfield music room that many of Oxford's musical soirées took place.

As soon as Margaret Deneke heard that Einstein was coming to Oxford, she invited three other musicians to come and to play string quartets with him. All three accepted, and they constituted a remarkable musical ensemble: apart from Einstein, they were eminent musical performers with close personal ties to some of the great names of nineteenth-century music, including Johannes Brahms, Clara Schumann, and Joseph Joachim. Marie Soldat played first violin in this string quartet; Einstein played second; Erna Schulz, viola; and Arthur Williams, cello; occasionally, they were joined by Margaret Deneke, playing piano. Soldat had in her youth been a protégée and friend of Brahms, and it was on his instigation that she became a student of the renowned violinist Joachim when she was sixteen. She became one of his star pupils, and her performances of the Brahms Violin Concerto won her the admiration of Brahms, who introduced her to his circle of friends, in particular the Wittgenstein family. Erna Schulz had been another of Joachim's star students and had enjoyed a brilliant career, and Williams, the cellist, had studied with Brahms's favorite cellist, Robert Hausmann, and had been a member of the celebrated Klingler Quartet.[20]

In her memoir, Margaret Deneke recalled her first meeting with Einstein:

In he came with short quick steps. He had a big head and a very lofty fore-head, a pale face, a shock of grey, untidy hair [. . .] [After dinner] he turned to Mother with an engaging smile: "You have provided a delightful meal; now the enjoyable part of our evening is over and we must get down to work on our instruments. Shall we play Mozart? Mozart is my first love—the supremest of the supreme—for playing Beethoven I must make something of an effort, but playing Mozart is the most marvelous experience in the world." The players started with a Mozart; under Marie Soldat's rich tone on her Guarnerius del Gesu,[21] the Professor's borrowed violin sounded starved and raucous, but his rhythm was impeccable. Before passing on to a Haydn there was an interval for a chat. Marie Soldat commented on the Professor's long violin fingers, tapering usefully at the tips. The professor said, "Yes I never have practiced and my playing is that sort of playing. . . ."[22]

The first meeting of the quartet took place at Gunfield on May 11, and Einstein enjoyed himself hugely. Still, music making was not the only reason that Einstein felt very much at home with the Denekes and enjoyed their company: since both sisters spoke German fluently, they were among those in Oxford—Lindemann was another—with whom Einstein could converse freely in German.

Einstein continued to work on the identification of field equations that were suitable for a unified theory, but did not report any success. Since he was without his usual secretarial help, he had to take care of his correspon-dence by himself. At a meeting with a group of Zionist students, he heard a talk about the Arab question, an issue that concerned him deeply. He won-dered whether he had convinced the speaker of his views. He also talked with student supporters of the League of Nations, and with those promoting pacifism. Apart from that, there was a procession of professors, func-tionaries, and dignitaries that he had to contend with. On a lighter side, word of Einstein's love of sailing had gotten around, and Douglas Roaf, a physics research student, took him sailing on the Thames in a skiff.[23]

Einstein visited Wadham College in order to meet with the eminent

Einstein, Elsa, and Margot in their Berlin apartment, ca. 1927. *Bildarchiv Preussischer Kulturbesitz (BPK)/ Art Resource.*

Einstein in his "turret study" in the Haberlandstrasse, Berlin. Newton's picture can be seen on the wall. *Bildarchiv Preussischer Kulturbesitz (BPK)/ Art Resource.*

Four of the ships that Einstein sailed on. *Anon. Public domain, Old Ship Pictures Galleries.*

Kitano Maru (1922)

Cap Polonio (1925)

BELGENLAND (1930)

DEUTSCHLAND (1931)

Einstein and Hayasi Myake (probably) on their way from Marseille to Japan. *Courtesy of the Leo Baeck Institute.*

Elsa and Einstein pose with Sir Herbert and Lady Samuel in front of "Samuel's Castle," Jerusalem, 1923. *Parliamentary Archives, London.*

Elsa an Einstein boarding the Belgenland at the beginning of their voyage to San Diego via New York,
1930. *Bildarchiv Preussischer Kulturbesitz (BPK)/ Art
Resource.*

Einstein and Elsa's reception in New York appears to
bewilder him, 1930. *Bildarchiv Preussischer Kulturbesitz (BPK)/ Art Resource.*

The Einsteins' home in rural Caputh, 1930. *Courtesy of the Leo Baeck Institute.*

Rabindranath Tagore and Einstein leaving his Caputh home, 1930. *Courtesy of the Leo Baeck Institute.*

Einstein with the jovial captain of the Portland, 1931. *Bundesarchiv.*

Einstein watches as the pilot assists Senora Wasserman in boarding a Junkers
seaplane. It was Einstein's first flight.
Buenos Aires, 1925. *Courtesy of the Leo
Baeck Institute.*

Einstein and two members of the ship's orchestra playing a piano trio aboard the Deutschland, 1932. *Bildarchiv Preussischer Kulturbesitz (BPK)/ Art Resource.*

Einstein surrounded by his welcoming committee in Montevideo, 1925. *Courtesy of the Leo Baeck Institute.*

Einstein relaxing with the philosopher Vaz Ferreira on a park bench in Montevideo, 1925. *Courtesy of the Leo Baeck Institute.*

Einstein with the physicist Max Tolman at Caltech, 1932. *Los Angeles Times.*

Einstein with the solar astronomer Charles St. John, 1932. *Bildarchiv Preussischer Kulturbesitz (BPK)/ Art Resource.*

Elsa and Einstein in a cactus garden, Palm Springs, California, 1932. *Courtesy of the Leo Baeck Institute.*

Elsa and Einstein at ease with their host, Samuel Untermeyer, at his home in Palm Springs, 1932. *Bundesarchiv.*

Einstein and Charlie Chaplin at the opening of *City Lights*, Los Angeles, 1931. *Courtesy of the Leo Baeck Institute.*

Einstein arm-in-arm with the pianist Leopold Goldowski and the composer Arnold Schönberg at a benefit concert in Carnegie Hall, 1934. *Bildarchiv Preussischer Kulturbesitz (BPK)/ Art Resource.*

Winston Churchill and Einstein in the garden of Churchill's home, Chartwell, Kent, 1933. *Courtesy of the Leo Baeck Institute.*

Einstein at the tiller of his sailboat, the Tümmler, 1930. *Courtesy of the Leo Baeck Institute.*

according to Einstein, and not religious.[24] At other times he went for a long walk with H. G. Fiedler, the professor of German literature, or watched the rowing regattas on the Thames; another time, he heard a picturesque, but outrageous, charlatan lecture about ancient Coptic music. Einstein described the speaker as 'a fat giant with red face,' wearing a blue monk's habit, who spoke, as if drunk, from a blue podium.

In the evening of May 21, before a festive dinner given by the dean of Christ Church, Einstein's servant laid out for him a fresh tuxedo shirt and collar—without having been asked, noted Einstein. At dinner, he was seated between 'two dragons, who spoke only English,' and while Einstein generally considered himself a nondrinker, he found on this occasion solace in the 'excellent port wine' and the ice cream. Once he was happily back in his comfortable lodgings, a great sense of contentment came over him, for the dean's dinner was the last of the obligatory 'ceremonial feedings' (*Paradefrass*) he was committed to attend. 'Hallelujah.'

On May 22 Einstein paid a visit to Ruskin College, which was accommodated in a modest house. The college, named after the artist, critic, and social reformer John Ruskin, was founded in 1899. Not formally associated with Oxford University, it was funded by public foundations and had strong links to British labor unions. It specialized in giving adults with few academic qualifications the opportunity to take university courses—a mission that Einstein approved of. He was much impressed by the forty students at that 'splendid institution,' and he learned that nineteen of its former students had become Members of Parliament. Later, he enjoyed a 'lovely and most agreeable' evening playing string quartets at the Pearces' home with Soldat, Schulz, and a young local cellist. Starting with Mozart and Haydn quartets, they were later joined by a lady violist with whom they played a Mozart string quintet. Einstein had such a good time that he did not get home until midnight.

May 26 was a busy day for Einstein. In the morning, he took part in the Encaenia, the tradition-laden academic ceremony at which academic degrees were awarded—including an honorary DSc to Einstein. The solemn occasion opened with an encomium (celebratory speech) in Latin that Einstein, for reasons unknown, considered somewhat inappropriate. Afterward, he presented the last of his three lectures at the Rhodes House, which he devoted to the mathematical methods for field theories. His talk failed to disturb the 'glorious slumber' of the dean, who sat in the first row, but the remainder of the audience was exceedingly well-behaved and friendly. After the lecture, he had a nap in Lindemann's lodgings and ate in at the college. In the evening, he met with a pacifist student group in a charming old house and was astonished by the political maturity of the students, lamenting how pitiful '*our* students' were, by comparison.

The weather was cold and rainy the next day. It was Whitsunday, and Einstein had lunch at the Pearces' home, where he first played a Dvořák quartet with his usual partners, then a Mozart violin and piano sonata with a professor at the music department, and finally the Double Concerto by Bach. After the music, Mrs. Pearce showed Einstein her 'fabulous garden' with its small hill and artificial lake. Blue spring flowers were in bloom, and Einstein believed that the garden owed its fairytale quality to the absence of visible boundaries, so that it became a world of its own.

Back at the college, he skipped dinner and lay down on a mat in front of the fireplace to rest—and to keep warm. But at 9:00 p.m. he went to meet a group of philosophers with whom he engaged in a question-and-answer session—a format of discussion that really appealed to Einstein—about space and time. He enjoyed the evening a lot. He mused that a comparison between this 'wonderful and incredibly pleasant' exchange of ideas with the tense discussion he had had with Philipp Lenard—the most vociferous foe of relativity theory—at the 1920 Congress of German Physicians and Scientists showed him how wretched Germany was.[25]

Lindemann took Einstein to the country estate of old Lady Fitzgerald the following day. Einstein admired the garden but saved his greatest enthusiasm for the drinking fountains for cows he saw there. He explained that when a cow pressed her muzzle against a metal plate, a valve was opened and water flowed into a small bowl—as long as the cow kept her muzzle in place. Einstein wondered whether toilet facilities for cows would be next, adding, 'long live civilization!'

In the evening, he again played chamber music at the Denekes' home, but this, their last session in Oxford, left Einstein discomfited because Soldat and Schulz squabbled incessantly, while feigning friendliness. He sighed, 'O those womenfolk!' (*O die Weibsen!*)

Einstein and Lindemann visited the Ashmolean Museum the next morning to view artifacts recently excavated in Crete. Einstein thought their character more Egyptian than Greek. In the afternoon, he met with a delegation from War Resisters International, and these 'admirable people' discussed with him the dissemination of his antiwar speeches in America. In their opinion, his speeches had been very effective. In the evening, Einstein resolved in his diary to work for this cause with all his energy.

BACK TO REALITY

Einstein's last day in Oxford was May 27. He had a conversation with a young philosopher from Saint John's College in the morning, and in the afternoon he met with a group of Quakers to discuss pacifism issues generally and the participation of Catholic clergy in the pacifist movement specifically. Lindemann then took Einstein away to watch the last of the college regattas on the Thames and to observe the happy tumult of spectators lining the riverbank.

The weather was lovely on the day Einstein left Oxford. Lindemann drove him to Southampton, where he boarded the *Hamburg*, a twenty-one-thousand-ton liner of the Hamburg America Line. Once on board, it did not take Einstein long to make contact with a kindred spirit among his fellow passengers: the mathematician Wilhelm Blaschke, a 'clever and agreeable Austrian,' who was on his way home after lecturing in America.[26] Blaschke told Einstein that many of Europe's learned heads were busily raking in the dollars in America the same way. Einstein did not miss this opportunity to mention Mayer's situation to him.

On Einstein's last day at sea, he visited the bridge. He learned from the captain that two days before, Auguste Piccard had taken off in a high-altitude balloon of his own design, and that he had been missing since. Piccard was a physicist who had constructed the pressurized gondola in order to study the earth's atmosphere and to investigate cosmic rays, whose nature was still a mystery. Einstein, who knew Piccard personally, expressed the hope that he would survive the experiment. Piccard did survive; having reached a record altitude of sixteen kilometers, his balloon landed safely on a glacier in the Austrian Alps.

The *Hamburg* had almost reached the port of disembarkation, Cuxhaven, at the mouth of the Elbe estuary (about a hundred miles upriver from Hamburg), when a dense fog suddenly descended and forced the ship to drop anchor. Einstein noted that the fog was caused by cold winds from the north encountering warm seawater. As evening fell, the fog grew thicker, and Einstein became increasingly annoyed with himself for having spent a small fortune on radio telegrams announcing his arrival. Now, he had to send additional telegrams to cancel the earlier messages because of the fog. 'Never again!' he swore. But he quickly regained his good humor, admitting to his diary that life on board was really very pleasant, as long as the food held out and as long as one had such a luxurious cabin to stay in and is 'gawked at like the orangutan in the zoo.'

*   *   *

That was the last diary entry Einstein made on this journey, but on Monday, June 15, two weeks after he had resumed his normal life in Berlin and Caputh, he was motivated to append a somber entry.

He recounted attending a meeting of the German Physical Society on the previous Friday, at which Arnold Sommerfeld, Einstein's friend and scientific associate, was awarded the Planck Medal for his contributions to quantum mechanics. Planck had given a speech that Einstein thought was very witty but less than benevolent. Later, Freundlich announced the result of his latest attempt to measure the deflection of starlight caused by the sun's gravitational field, the experiment on which he had labored for the past seventeen years. His latest measurements yielded a deflection that was greater than predicted by relativity, but the experimental uncertainty was so large that it failed to clarify the situation. But this was hardly a matter of concern for Einstein: his confidence in relativity was total, and his personal relationship with Freundlich had cooled considerably over the years.[27] He recalled going sailing 'with Dima' (possibly, Margarete Lebach) on Sunday in his beloved sailboat, the twenty-one-foot *Tümmler* (*Porpoise*). Today he was meeting with a delegation of pacifists in the morning to discuss various issues related to the refusal of military service. Later, it was back to physics and a conference with Mayer on the integration of the latest field equations.

The deteriorating political situation in Germany weighed heavily on Einstein's mind. He noted that one could no longer purchase gold certificates in Germany and that a mood of panic prevailed in Berlin. Rumors of a putsch were in the air, and Einstein believed that disaster was imminent; the reactionary forces appeared ready to strike.

Memories of the contemplative college life, walks in the English countryside, and boat races on the Thames were receding swiftly.

7.

Return to Pasadena (1931–1932)

Einstein stayed at his country retreat in Caputh for the remainder of the summer of 1931. In pursuit of his struggle against militarism, he advocated the refusal of military service and signed numerous antiwar petitions and proclamations. He wrote personal appeals to kings and presidents on behalf of draft resisters who were under arrest in Bulgaria, Czechoslovakia, Poland, Serbia, and elsewhere.[1] As Germany drifted inexorably toward becoming a Nazi dictatorship, Einstein was coming to terms with the likelihood that he would have to leave Berlin. He was in negotiation with Caltech regarding an academic appointment, and while the negotiations continued, Einstein prepared to spend another term there—as had been agreed the year before.

ACROSS THE ATLANTIC

Einstein and Elsa left Berlin on November 14, 1931, and in the next two weeks they visited their many friends in Belgium and the Netherlands before reaching Antwerp, their embarkation port. At the railway station in Antwerp, they were met by Herr Geissler, an agent of the Hamburg America Line, who as a boy had been a classmate of Einstein's in Munich. He escorted the couple to the harbor and the dock where their ship, the MS *Portland*, was tied up.[2] The *Portland* would take them all the way to Los Angeles by way of the Panama Canal—the route Einstein had chosen evidently because it offered more days at sea and because it steered clear of New York. The ship remained in port until late evening of the following day (December 2), and Einstein grumbled in his diary that in the interim he had been somewhat exasperated by the protracted visits of unnamed family members (*Mishpoche*), who came aboard to see him off.

Einstein must have been in an unusually effusive frame of mind when he made the first entry in his new diary, the next day, for he recorded almost verbatim a bizarre story, told to him by Geissler, who since his school days had turned into a 'worthy, somewhat Prussianized Bavarian,' a serious chess player, and a drinker. The story was this: As a young boy, Geissler despised his family's vegetable soup (*Französische Suppe*) but was strictly forbidden to leave any food on his plate. Once, when, after gulping down a few spoonfuls of vegetable soup at the midday family meal, he gave up in despair, his father, a very 'energetic army officer,' left the table to fetch his pistol. He placed the pistol on the table and told his son, "Eat the soup, or I will shoot you." Young Geissler forced a few more spoonfuls of the hated brew down his throat before he capitulated, unbuttoned his little waistcoat, and very sincerely said to his father: "shoot!"

Although the *Portland* was relatively small (6,700 tons), Einstein found her to be quite respectable, and he took an immediate liking to her jovial captain. But in his diary he expressed sympathy for the *Portland*'s shareholders because on this crossing, the ship carried only ten cabin passengers—she had accommodations for fifty—and hardly any freight. He and Elsa occupied adjoining cabins, and, at meal times, they sat at their own little table. During his first two days at sea, Einstein proved to himself that all of Euclid's geometrical constructions could be performed with a compass alone; that is, without a ruler.[3] What motivated him in this undertaking was a conversation with the physicist 'Tania' (Tatiana) Ehrenfest, whom he had visited in Leiden the week before;[4] she had insisted that a ruler was essential. When the *Portland* reached Cristobal, Panama, her first port of call, Einstein mailed the proof to her.

On his way to Antwerp, Einstein had spent several days at Leiden University, where he had many friends and colleagues and held a special visiting professorship. In his diary, he recalled delivering a lecture at the university just one hour after arriving in Leiden, and a colloquium the following evening. He and Tania Ehrenfest had studied the new five-dimensional theory of Veblen and Hoffmann[5] together; it had seemed quite artificial to them, and Einstein wondered now what Veblen and Hoffmann might make of *his* latest theory! He had also gotten together (in Utrecht) with the astronomer Willem Julius, with whom he kept up a lively correspondence for many years. Einstein had taken a long walk with Adriaan Fokker, his first postdoc,[6] but he had skipped the lecture by the visiting Austrian physicist Adolf Smekal[7] in order to visit the Dutch parliament in The Hague instead. Finally, Einstein recalled the very pleasant (*gemütlicher*) evening with the Ehrenfests at the Fokkers' home, and he avowed that 'it was really lovely in Leiden.'

Einstein recalled with particular pleasure his tour of Leiden's natural history museum, accompanied by the outstanding instrument maker Claude Crommelin, who was also the museum's cofounder. Crommelin demonstrated a steam-driven electrical generator to Einstein and showed him the primitive but remarkably powerful microscopes that enabled van Leeuwenhoek to make his dazzling biological discoveries in the seventeenth century. (With hundreds of such microscopes that he built himself, van Leeuwenhoek observed and described for the first time red blood cells, animal sperm, living bacteria, and many other microorganisms.[8]) To top off their museum tour, Crommelin produced a postcard Einstein had written to Professor Kamerlingh-Onnes in April 1901 to apply for an assistantship in his laboratory.[9] The card harked back to the time when, after graduating at the ETH, Einstein was desperately searching for an academic position, having failed to obtain one in Zurich. How much things had changed!

By December 5, the *Portland* had cleared the English Channel. Although the weather was getting warmer, it rained continuously and the sea was turbulent. Einstein settled into life on board and praised its 'enviable tranquility.' He had brought a small library with him and began reading his way through it, opening first the recently published third (and final) volume of Egon Friedell's monumental cultural history.[10] Einstein was full of admiration for Friedell's erudition, his fine intellect, and his beautiful language, and he expressed indebtedness to his friend Toni Mendel for having drawn his attention to Friedell. Einstein also read in Max Born's recent book on quantum mechanics, as well as a volume of the legends of South American natives that had been collected by the ethnologist Koch-Grünberg in his years of living among the Pemon Indians.[11]

Einstein's taste in reading matter was remarkably eclectic. He turned next to a compilation of 'Chinese wisdom' that had recently appeared in German translation. After he read it, he confessed to having understood none of it. He had, on the other hand, understood Carl Jung's commentary but found it worthless 'drivel'—empty words without a clear line of thought.[12] If a *Psychaster* is needed, Einstein concluded, then let it be Freud, whom he admired for his concise writing and his original, if sometimes too wideranging, spirit—although he did not believe Freud's theories.

The diary entry of December 6 is particularly noteworthy, for it announces Einstein's decision to give up his position in Berlin; Einstein wrote that he would now become a 'migratory bird' for the rest of his life. In spite of anti-Semitism and political turmoil, many strong personal and cultural bonds still tied him to Berlin. He mused that the gulls accompanying the *Portland* were now his new colleagues, then added ruefully that their skill and geographical knowledge far exceeded his. As if to underscore his resolve, Einstein began to study English, only to conclude that 'it just won't stick to my ancient skull.'

The sea had become very agitated, and Einstein marveled at the vulnerability of man when confronted with nature's fury. Because the *Portland* carried so little cargo and no ballast, her propeller left the water every few seconds, causing the engines to race, only to come to a momentary complete stop when the propeller resubmerged. Only after the ship had passed the largest of the Azores did the weather improve so dramatically that Einstein spent the whole day on deck relishing the 'wonderful sunshine.' He chatted with one of the ship's officers, and, for a while, an excellent idea for making progress on his current theory occurred to him—that is, before he discovered that it was worthless.

On December 9, Einstein descended into the bowels of the *Portland* to inspect the engine room and the ship's diesel engines. He took an immediate liking to the intelligent chief engineer, who would be his companion on many occasions during the voyage. His handful of fellow passengers, in contrast, seemed particularly ordinary to Einstein. The finest person on board, he decided, was the tall ship's steward, whose distinguished demeanor gave him the appearance of a tax assessor or a foreign affairs officer. After just one week on board, Einstein had recaptured the contentment he usually felt at sea, and gave voice to it: 'Life on board ship is lovely.'

The weather changed again, and suddenly, the *Portland* found herself in the midst of a full-fledged Atlantic gale. Huge waves buffeted the lightly laden ship, and the glassware flew off the mess tables. Einstein, who was by now a seasoned seafarer, had never experienced a storm of such vehemence. The howling wind made a deafening noise, and everything that was not firmly fastened down went tumbling. Einstein, meanwhile, thought that the storm was 'indescribably magnificent,' especially when the sun illuminated the raging sea. It gave him the feeling of being wholly dissolved in nature, and he sensed the insignificance of the individual—a thought that cheered him even more than usual.

By December 12, the sea had become calm again. The weather was warm but not yet tropical, and flying fishes made an appearance. Tropical rain clouds came into view and reminded Einstein of the illustration in *Der Struwwelpeter* (the children's book) that shows "Flying Robert" being taken aloft by his umbrella during a rainstorm.[13] Einstein felt pleased with himself for having battled his paunch successfully on this voyage. He credited his achievement to his careful eating habits and to the seawater baths he took each day. That, he concluded, and finding himself so remote from all humankind were the reasons he was feeling so exceedingly well.

The deplorable political situation in Germany was, however, never far from Einstein's mind. Upon hearing the latest emergency decrees issued by Chancellor Brüning[14] and other news received over the ship's shortwave radio, he concluded that property ownership was becoming increasingly problematic in Germany. The economic pressure that France and the United States were exerting on Germany forced the country, in his view, into a sort of involuntary state capitalism (or, if you like, socialism), which would have consequences for those nations. Einstein noted of those consequences, 'They really have begun already.'

PANAMA AND HONDURAS

In the midst of a great heat wave, the *Portland* arrived in Cristobal, the Atlantic entry port to the Panama Canal, in the evening of December 17. The local German consul invited Einstein and Elsa, along with the captain, to dinner at his home. On their way back to the ship, they had an opportunity to observe Cristobal's nightlife. Einstein liked the colorful crowds milling in the streets and was intrigued by a row of small, lighted shacks, with a nattily dressed girl sitting in front of each.

Following an oppressively hot night aboard, Einstein and Elsa were again invited ashore in the morning. After being driven over exceedingly bumpy dirt roads, they reached a luxuriant rain forest, where Einstein was deeply impressed by the wonderful trees and flowers, the exotic butterflies, and the twitter of so many strange and unfamiliar birds. Einstein and Elsa were then shown the newest, electrically operated three-stage lock of the Panama Canal. While returning to the ship, they passed a shallow river in which two alligators were seen to lounge.

It was tremendously hot aboard the *Portland* that afternoon, and Einstein was on deck observing the picturesque activity in the harbor. Some unfamiliar birds soared 'wonderfully' overhead; they resembled swallows, but they were larger than gulls, with narrow, arched wings and a forked tail. The much-traveled chief engineer joined Einstein on deck, and the two chatted about the current state of affairs in America.

Early in the morning (December 19) the *Portland* entered the Canal. As she passed through the Gatun lock, the processed prints of photos that the chief engineer had taken the day before were passed onto the ship. It was extremely hot, and Einstein fell asleep for several hours during the passage through the Canal; as a result, he missed the ship's exit. When he awoke, it was evening and a cool breeze was blowing on deck. The night was hazy, and a colorless halo surrounded the moon. Far in the distance, lightning flashed silently. The ship had reached the southernmost point of her voyage.

By following the coast of Central America in a westerly direction, the *Portland* reached Puerto Moios, a 'banana port of a large North American fruit company,' where she docked to pick up a cargo of bananas. Einstein and Elsa accepted an invitation to come ashore and to inspect the enormous banana plantation, ten square miles in area, which was said to provide employment for 3,500 workers. On their tour, they were accompanied by the chief engineer and a local physician, and they traveled in a Ford automobile that had been modified to run on rails. At first, the party visited the vast banana fields and then came to a magnificent rain forest with an 'unimaginable richness of shapes.' Acting as their guide, the physician explained the American approach to colonization, claiming that the company provided first-rate medical care and hygienic living conditions for all its workers. Einstein and Elsa later met local residents, who talked to them of the issues that were uppermost on their minds: the newly available drug, Plasmocin, the first synthetic anti-malarial drug; and their worries about the dreaded soil-borne 'Panama disease' that infects banana and plantain plants. They also talked to Einstein and Elsa about the character of the native Indians, who, they said, resembled Eskimos in their appearance and were under threat of extinction. The Indians were apparently doomed either to work on the plantation or to starve, if they chose to retreat into the mountains. Einstein came to the conclusion that in the tropics, the fight for survival was even more brutal than in Europe—for people, for animals, and for plants. Every organism had to struggle against other organisms, with the corporation at the apex, itself driven by anonymous shareholders hungering for dividends. What an inspiring scenario, reflected Einstein; alas, there was no way out: 'eat or be eaten, without aim or end.'

In the morning of December 22, the *Portland* arrived in the Bay of Fonseca, Honduras's only access to the Pacific Ocean. She dropped anchor in the blue water of the bay among the many green volcanic islands and near the Isla del Tigre (Tiger Island). A Frenchman and a German woman from Bremen came aboard to chat with Einstein and Elsa. In the course of their conversation, Einstein learned that only a few white families lived there and that all of the natives were suffering from malaria; most had syphilis too. At noon, the *Portland* raised her anchor and departed. The temperature in the shade was 36°C (97°F). Einstein was on deck as the ship made her way out to sea, and he was thrilled by the 'indescribably picturesque scenery' ashore. His attention was caught by a huge volcano far in the distance, with a smoke cloud hovering over its summit. It remained visible for a long time.

At one in the morning, the *Portland* arrived at the tiny El Salvadorian port of La Libertad. Einstein's presence on the ship had again become known ashore, and once more, a man and a woman came aboard and wished to speak with Einstein. But when Einstein heard that the man had served as secretary to Wilhelm II during the kaiser's exile in Doorn, he declined to meet with the pair—much to the expressed annoyance of Elsa, who would have enjoyed an interesting diversion.

Underway again later in the morning, the ship motored west along the Guatemalan coast, past a chain of tall volcanoes on the mainland. Einstein enjoyed the cooler evening weather and the spectacular sunsets, but the news from Germany continued to be disheartening: the huge Borsig machine works in Berlin, where locomotives had been built since 1840, had stopped paying its bills and was on the verge of bankruptcy. Einstein sensed disaster looming.

Einstein finished reading the final volume of Friedell's *Cultural History*, and though the book ended with some 'blooming nonsense' about the exact sciences, what did it matter? It made the excellent rest of the book all the more entertaining. Einstein next took up the recently published *History of the Jews*, whose author, Josef Kastein, argued that Jews were not sufficiently aware of their past and saw themselves too much through the eyes of others.[15] Einstein, although hardly a Zionist, found the book inspiring.

On December 24, the *Portland* was sailing along the Mexican coast, and though it was hot and humid, a cooling breeze blew all the time. In the evening, there was a Christmas party, complete with a sparkly Christmas tree, carol singing, and a jovial dinner at which the passengers enjoyed caviar and turkey. After the meal, 'the better sort of passengers' were invited to mingle with the captain and the ship's officers in the salon. For a while, Einstein thought he had discovered a way of deriving the Schrödinger equation from his latest field theory—but the idea came to naught. 'A disappointment!'

The traditional festive farewell dinner took place three days later. The captain gave a simple and straightforward speech that won Einstein's approval. The next day, when they were already off the coast of California, the ship ran into a powerful storm that produced wave trains of greater length than Einstein had ever seen before. Kastein's book continued to affect him deeply. He wondered *for what* his Jewish ancestors had let themselves be slaughtered by the hundreds of thousands, adding that it must not be forgotten that the Jews have been—and are—among the 'main emissaries of the cosmopolitan idea.'

PASADENA ENCORE

When the *Portland* arrived in Los Angeles late in the evening of December 30, Einstein found the city was a 'sea of lights' and its harbor full of warships. He had requested that there be no welcoming ceremony on his arrival, but he agreed to a morning press conference on board. The usual questions were submitted to him in writing, and he answered them with Richard Tolman at his side acting as his interpreter. With that chore behind them, Tolman drove Einstein and Elsa to Arthur Fleming's home in Pasadena, where they stayed for their first five days ashore after spending a month afloat. Since the Fleming household was run by Arthur's sister, Elsa asked her to ensure that Einstein's food was not spicy—in view of his culinary experience at the Flemings' the year before (see chapter 5). Einstein enjoyed being back in the Flemings' wonderful, steeply sloping garden. He marveled at the variety of birds he saw there and was particularly intrigued by the hovering hummingbirds.

On January 6, Einstein and Elsa moved from the Flemings' home to a suite in the Athenaeum, which served as Caltech's faculty club and as a venue for meetings. It would be the Einsteins' residence for the remainder of their stay in Pasadena, even though Elsa would have much preferred a bungalow with a garden to tend and a kitchen to cook in, as in the year before—according to a *Los Angeles Times* reporter.[16] During his first few days at Caltech, Einstein attended seminars on the ionization of noble gases and on relativistic thermo-dynamics. He also heard the latest rumor circulating among the Caltech physi-cists: that Harold Urey and his associates at Columbia University had discov-ered an isotope of hydrogen: deuterium (heavy hydrogen).[17]

But physics was not all that occupied Einstein in Pasadena; he com-mented on the fact that 10 percent of the town's population was unemployed at the time. On New Year's Day, he and Elsa again watched the Tournament of Roses parade and were serenaded by a band that included girls playing trombone, something Einstein had not encountered before. The next day, he was photographed standing next to a parade float that celebrated the youth of Germany. He was continually surprised by the enormous sympathy for Germany and the hatred of France that he encountered in Pasadena. He real-ized, on the other hand, that the political views of the people he had encoun-tered were remarkably ill-informed and immature.

Einstein enjoyed staying at the Athenaeum, where he often ran into vis-iting academics he knew. Among them was Franz Simon, the physical chemist, known to Einstein from Berlin, and the Viennese philosopher Moritz Schlick, a 'fine head' according to Einstein, with whom he debated the pros and cons of positivism.[18] The Dutch astronomer Willem de Sitter was another visitor, and Einstein collaborated with him on a short article in which they derived a relationship between the mass density of the universe and its rate of expansion.[19] In his search for a unified field theory, Einstein made no progress, however. An enormous amount of correspondence awaited his attention, and the task of dealing with it was not made easier by the 'slim and exceptionally dumb secretary' who had been assigned to him.

The latest economic and political events in Germany caused Einstein deep concern, and he foresaw dire times ahead, but he was also concerned by troublesome developments in America, where the economic depression was deepening, even as President Hoover refused to meet with a delegation of unemployed workers. Tom Mooney, a union organizer who had been con-victed of murder under highly dubious circumstances, was a celebrated martyr of organized labor; a delegation of unemployed workers repeatedly appealed to Einstein to visit him in prison.[20] Einstein, who had promised Millikan he would stay out of American politics while at Caltech, very wisely refused to 'take part in these theatricals.' He had, on the other hand, already written to the governor of California on Mooney's behalf.

On January 10, Einstein and Elsa traveled to Palm Springs to spend the weekend at Samuel Untermeyer's luxurious estate, where Einstein greatly enjoyed the glorious desert sunshine—as in the year before. One day, during lunch at a Palm Springs hotel, he had the unexpected pleasure of being invited to join the Irish string trio that was performing at that establishment. He was lent a violin and happily joined the three other musicians in pieces by Bach, Mozart, Beethoven, and Handel.

Einstein was also introduced to a professional psychic in Palm Springs. Though it is not known whether he believed her powers were genuine, he was evidently intrigued by her perceived 'talent': having just met Elsa, the psychic told her that she was powerfully (and painfully) attached to her mother; that she had two daughters, one with blue eyes, the other with brown eyes; and that the latter was unhappy and was suffering from a stomach ailment. Einstein described the psychic as a beautiful and kindhearted twenty-four-year-old woman who was carefree, adventurous, and healthy. Strangers had supposedly discovered her abilities very early, and she was now earning a fabulous income.

Einstein did not put his work aside in Palm Springs. In his diary, he wrote down a set of equations that were a necessary corollary to the field equations he had developed at the beginning of the voyage. The new corollary seemed sufficiently important for him to send it to his (much missed) 'Mayerchen' (Walther Mayer) in Berlin, as soon as he returned to Pasadena.

Having settled effortlessly into the academic life at Caltech, Einstein attended numerous lectures with topics that reflect the broad range of his interests. He heard Jacob Schurman, a former US ambassador to Germany, discuss the present situation in Germany, but his superficial, mocking account of German politics earned Einstein's disdain.[21] Einstein also listened to Millikan's lecture on cosmic rays, but while he thought the experiments were beautiful, he was puzzled by them and considered the theory weak. Later, Einstein had an opportunity to inspect a Wilson cloud chamber, the device that became a tremendously valuable tool in nuclear physics and cosmic ray research. The cloud chamber made it possible to visualize the tracks produced by cosmic ray particles in the presence of strong magnetic fields.[22] Einstein, who had only limited interest in nuclear physics, was left utterly mystified by Millikan's data. In his own, well-received colloquium

(January 19), Einstein discussed the current state of his work on a unified field theory. The next day, he visited Jesse DuMond's laboratory, where, ever the connoisseur of ingenious scientific instrumentation, he saw the 'wonderful' crystal spectrometer DuMond had built for determining the wavelengths of X-rays.[23] The crystal spectrometer, along with the cloud chamber, played a crucial role in establishing the Compton Effect, widely considered to constitute the most convincing evidence for the quantum nature of radiation. On a visit to the laboratory of Alexander Götz, a condensed matter physicist, Einstein was shown evidence for large complexes and a glassy phase near the melting point of metallic crystals—a result that had a bearing on Einstein's quantum theory of solids.

But there was also time to pursue interests other than those scientific. Whenever he could, Einstein liked to sit on his little balcony, soaking up the sun 'like a crocodile,' and every day, alone in his room, he played the violin. Occasionally, he joined others for chamber music: when Götz's wife turned out to be a good pianist, Einstein played a 'very beautiful Vivaldi sonata' with her, and he was left 'exceptionally content' following a wonderful evening of playing Bach, Mozart, and Schumann at Lili Petschnikoff's home in Hollywood. Petschnikoff and the Einsteins had met the year before; she was born in Chicago and spoke German fluently, for she had studied in Germany. She had been a celebrated violin soloist in her younger days, mostly before the First World War, and had played in concerts all over Europe. In her memoir, Petschnikoff recounts a musical evening with Einstein, commenting on his 'fleet fingers' and praising his good tone; she notes that when they played a duet, not even the most intricate Bach fugues flustered Einstein. While Einstein and Petschnikoff played, Elsa was busy with her sewing basket. As she later explained, she had mended her older stockings, which she planned to present to the maids at the Athenaeum when they left Pasadena. When Petschnikoff asked Einstein his views on marriage, he said that when he and Elsa got married, he had told her that he would take care of all major issues, while she would attend to the minor matters; so far, he added, they had all been minor.[24]

Galinka Ehrenfest, the twenty-year-old daughter of Paul and Tatiana Ehrenfest, showed up in Pasadena on January 17. The Einsteins were happy to act as her hosts and took her first to a puppet theater and then to a Mexican market in Los Angeles. They were all delighted by the spectacle of the carefree and colorful Mexican crowd milling about in the middle of a modern metropolis—a scene that brought *Paradise Lost* to Einstein's mind. Later Einstein enjoyed another exceedingly satisfying musical evening at Petschnikoff's home playing Bach, Mozart, and Schumann.

The mathematician Oswald Veblen turned up as another academic visitor at Caltech.[25] On January 22, he came to see Einstein in his suite at the Athenaeum and the two men chatted while sitting on the little balcony. Einstein was left with the impression that Veblen was very clever, but a bit full of himself. In his reading, Einstein turned to philosophy and began José Ortega's *The Revolt of the Masses*. Although Einstein found Ortega's style pretentious, he praised him for his intelligence and for refuting Spengler's nonsensical writings about history.[26]

In contrast to the year before, Einstein was not troubled by excessive publicity on his second visit, and he thoroughly enjoyed the comfortable lifestyle in Pasadena. He ate most meals, except dinner, in his room, and he gave no public speeches, with two exceptions: on January 18, he addressed a disarmament conference organized by Quakers in the nearby town of Whittier, and he spoke again, very reluctantly, at a huge Caltech banquet one week later.

Einstein also heard a lecture by Robert Oppenheimer, in which he discussed neutrons—neutral particles with the same mass as protons whose existence he was proposing on purely theoretical grounds. Astonishingly, James Chadwick discovered neutrons in a simple but ingenious experiment just one month later.[27] The confusing state of theoretical nuclear physics mystified Einstein, and he felt sure that these recent developments represented a cul-de-sac. That evening, he and Elsa were invited to the home of the German consul. The picture of President Hindenburg that hung on the wall reminded Einstein of a family photo of a Mecklenburg innkeeper. They watched the consul's little brats perform a classical ballet—to the accompaniment of Handel's *Largo*. 'God save us! Good thing He did not see it.'

After a week in which he made no entries, Einstein took pity on his 'neglected diary' on February 3 and brought it up to date. The weather had been cold and miserable all week, and it rained a lot. Both he and, particularly, Elsa had suffered from severe colds. Einstein had spent a lot of time discussing Veblen's theory of electricity with him and muttered that mathematicians like Veblen were 'empiricists with pencil and paper.' All the same, Einstein had now taken a liking to Veblen: he was a nice, modest fellow—as long as you 'get him *alone*—as is so often the case!' The day before, Einstein had taken part in a small luncheon with the objective of encouraging a wealthy lady to provide funds for a laboratory assistant of Jesse DuMond—'a fine physicist with a pretty French wife,' according to Einstein. There was a concert of the Los Angeles Symphony Orchestra four days later, at which Einstein and Elsa heard an excellent performance of the symphony *Israel* by Ernst Bloch, with Artur Rodzinski conducting that 'wonderful work.'

The following Sunday (February 7), the rain poured in buckets, giving Einstein an opportunity to bring his diary up to date once again. He had just returned from a luncheon in the home of Henry Robinson, a wealthy banker and major benefactor of Caltech; President Hoover's son had been among the guests—'a boring affair, but a good cigar,' observed Einstein. Much to the annoyance of the local bigwigs (*Grosskopfeten*), Einstein went to a 'negro church' in the afternoon to take part in a memorial service for the late Julius Rosenwald, who had been an unstinting philanthropist and a champion of education for African Americans for many years.[28] Einstein was the guest of honor at the service, and in his brief address he pleaded for racial tolerance and world peace. A rabbi followed him with a 'pretty speech,' and then the congregation sang a song of liberation, which was intended to raise their awareness. But in other respects, the service seemed 'peculiarly raw' to Einstein. To the bemusement of Einstein and few others, a black parishioner then sang a German song whose opening line is "May God protect you" (*Behüt' dich Gott*), but which was actually a popular song that deals with the breakup of two lovers.[29]

On the Friday before that, Einstein had attended a Shabbat service at the local synagogue and found that in spite of the exceedingly naïve prayers, the service was, on the whole, moving and poetic—particularly the *Shalom Ale-ichem*, the traditional song for welcoming the Shabbat.

On January 25, Einstein took part in a large banquet at the Athenaeum in support of world disarmament. Rufus von KleinSmid, the ambitious past president of the University of Southern California, told some tasteless jokes, but a retired marine officer gave a good, factual speech. Einstein also spoke, but he felt that it was, unfortunately, a case of pearls cast before swine. The entire affair left him feeling that the local affluent class would support any good cause merely to ward off boredom, and that this should not be confused with genuine commitment. He bemoaned the fact that these were 'the same people who call the tune in this pitiable world of ours.'

Such were Einstein's thoughts as he sat in a rattan chair on his balcony, savoring the warm California sun. Looking out toward Mount Wilson, he noted that a light snow had fallen overnight.

In his next and last entry (the dating is uncertain), Einstein recorded his activities on one particular day in Pasadena. Veblen, now well-liked by Einstein, called on him in the morning, and the two walked together to the astronomy seminar. Along the way, Einstein told Veblen of his futile efforts to obtain funds that would allow Mayer to join him at Caltech. Veblen, unsurprised, told Einstein that at Caltech, it was quite impossible to raise funds for anyone who was not already famous. They arrived late at the seminar but were in time to hear a 'wonderful lecture' by Walter Adams of the Mount Wilson Observatory in which he described an ingenious spectroscopic method for determining the distance of a star (spectroscopic parallax): if the star's mass was known (e.g., for Cepheids), and its temperature was estimated from its spectrum, its distance could be estimated from its apparent brightness. That afternoon, Einstein visited his violin partner, Lili Petschnikoff, who blurted out to him the story and strange fate of her former husband, whom she had divorced fifteen years earlier because of his multiple infidelities.

Afterward, Einstein went to hear another lecture by the diplomat Jacob Schurman, who discussed the current Manchurian crisis that had been precipitated by the Japanese invasion (in September 1931). Einstein thought that Schurman's account was intelligent enough, but quite amoral—'the source of all misfortune.' Finally, at the end of the day, Einstein was visited by a young astronomer who wished to test general relativity experimentally and sought Einstein's advice. Drawing on what he had heard in Adams's lecture that morning, Einstein suggested that the young man should determine the radius of Sirius B from its apparent brightness and known mass and relate it to its gravitational redshift.

At this point the diary peters out, except for a few jottings of impressions Einstein had gained in Pasadena. He ruminated that Pasadena seemed to be crawling with musical *Wunderkinder*, speculating as to whether that phenomenon was related to the abundant sunshine and the early maturation of children. As adults, on the other hand, the locals showed little originality and reminded him of scentless flowers.

Einstein's final comment was devoted to a young Pasadena woman whom he names only as Denise. She had told Einstein that she wanted to have an amorous adventure with him—for the sake of the publicity it would bring her. And in her latest effort to seduce him, Denise was plying him with all kinds of delicacies . . . !

*   *   *

The day Einstein and Elsa left Pasadena to return home (March 4), he consented to a final, good-natured press conference. After fielding the usual, mostly inane, questions, he was asked whether he was beginning to find it easier to talk to reporters. He replied that there was a German proverb that says that anyone can get used to being hanged.

With that chore behind him, he and Elsa boarded the MS *San Francisco* in Los Angeles to begin their long journey home.[30] They were seen off by Max Tolman, Einstein's frequent escort and interpreter in Pasadena, along with Paul Epstein and his wife, several German consular officials, and officials from Hapag. The *San Francisco*, similar in size and in accommodations to the *Portland*, made her way south, traversed the Panama Canal, crossed the Atlantic, and, after twenty-five days at sea, landed Einstein and Elsa safely in Hamburg.

8.

Oxford, Pasadena, and Last Days in Europe (1932–1933)

Einstein was back in Berlin and his rural retreat in Caputh by the end of March 1932, and he quickly resumed his busy academic, social, and political life. He attended weekly meetings of the Prussian Academy and submitted a manuscript to the Proceedings of the Academy in which he reported on his latest efforts to find a unified field theory. He also took up his political activities again and cosigned a manifesto warning that Germany was in danger of becoming a Fascist state. His face appeared on posters calling on the Social Democrat and Communist parties to form a united anti-Fascist front; but this failed to materialize, chiefly because the Communists saw the Social Democrats as a greater threat than the Nazis.

In the July election, the Nazi party won 37 percent of the delegates, making it the largest party in the Reichstag. President Hindenburg appointed as chancellor Franz von Papen, who, like most of his cabinet, was a member of the Prussian aristocracy. Papen's "government of the barons" dissolved the Reichstag, governed by decree, and employed the army (*Reichswehr*) to force the Social Democratic government of Prussia out of office.

Philipp Frank was visiting his friend Einstein in Caputh while these dramatic events were taking place, and in the course of their conversation, Einstein made the prediction that the people would turn to Hitler to protect them against the rule of the *Junkers* and the army.[1]

This time, Einstein's political instincts were right: it was the beginning of the end for the Weimar Republic. True to his resolve to become a 'migratory bird,' Einstein set out to explore suitable nesting sites outside Germany. While he had agreed to return to Caltech for a third time in the coming winter term, his negotiations with Millikan regarding a longer-term appointment remained inconclusive. A position for Mayer remained the stumbling block.

SECOND SOJOURN IN OXFORD

In April 1932, only two weeks after returning from Pasadena, Einstein was on the road again. He traveled to Cambridge to give several lectures and to confer with Arthur Eddington. From there, he continued to Oxford and took up residence in Christ Church, the college at which he had resided previously. During his first visit to Oxford, in 1931, the university had awarded him a research studentship entitling him to a yearly stipend of £400 with lodgings and dining rights for a period of five years. This seemed a good time to make use of his studentship.

Einstein must also have been looking forward to seeing his musical friends the Denekes again. And Margaret, for her part, wasted no time making contact with Einstein. On April 30, she came to Christ Church College with a large bouquet of garden flowers, which she handed to the college porter along with a note to Einstein. Two minutes later, the porter reappeared with a "smiling expression and a knowing look" as he descended a set of stairs, for he was closely followed by Einstein, who was rubbing the back of his head as he tried to read Margaret's note. According to Margaret, his flowing gray hair was longer than it had been the year before. He shook her hand warmly and asked how everyone was. Margaret told him that Marie Soldat and the other members of the quartet would be arriving on May 4, and she invited him to dinner at Gunfield, the Denekes' home, for the following evening.

These and many other minutiae have survived because Margaret Deneke was in the habit of recording each of her encounters with Einstein in a notebook soon after it took place. She wrote down his comments, often verbatim in flawless German, and her own remarks in English. The following excerpts from her notes reveal Einstein's ease in the Denekes' company.[2]

Although the morning of May 1 had been glorious, it rained in the afternoon when Helena and Margaret Deneke drove to Christ Church to pick up their dinner guest. If it would not offend Helena, said Einstein, he would prefer to walk, and so Margaret accompanied him on the sidewalk, while Helena followed behind in the car. He told Margaret how much the various types of people he saw on the streets of Oxford interested him, and how different they were from people at home. Einstein knew that Margaret had recently visited Albert Schweitzer at his hospital in Lambaréné (today Gabon, West Africa), having received a postcard from that place. He asked her whether Albert Schweitzer was a pacifist, and while Margaret was unable to provide an answer, she told Einstein that the most committed pacifist she knew was the violinist Adolf Busch—something that interested Einstein enormously. When they arrived at Gunfield, Margaret and Helena showed Einstein the violin they had borrowed for his use. He took it out of its case, tried it out, and pronounced it to be the worst instrument he had ever held in his hand. He claimed that the violin was "totally hoarse" but suggested playing a little Mozart, all the same, if Margaret would kindly play the piano very softly. After they played some Mozart sonatas, Einstein said, "Well, the sound is not beautiful, but you can't keep Mozart down—such a wealth of beauty!"

Einstein was the Denekes' guest for dinner and for music making on several other occasions, including on May 12. It rained again that day, but Einstein once more chose to walk to Gunfield, a distance of almost two miles. He carried an umbrella and wore a big mackintosh and his large, black, well-worn felt hat. When Schweitzer's famous book about Johann Sebastian Bach was mentioned, Einstein remarked that he had not read it and that, quite generally, he had no interest in discussions of art—that it was the passion itself he loved. Margaret mentioned that her family was not wealthy but always lived simply enough to create the feeling of wealth and the capacity of giving to others. That does constitute real freedom, agreed Einstein; you were truly human only when you discover that you desire nothing for yourself and are indifferent to social standing and such matters. But to be genuinely free, it was essential to be financially independent, for one could hardly be inwardly free while living in servitude to others. It gave him great satisfaction, said Einstein, that he was able to change his surroundings in these difficult times. Berlin was a very depressing place now, but when one was somewhere else, one saw the world very differently.

The conversation turned to diaries, and Einstein remarked that he had kept a diary on his recent American tour, but the most interesting things he had experienced he had never written down. His experiences during the war years in Berlin could have made a really worthwhile book, but if one doesn't record them while they are still scalding hot, he said, the details get lost. Margaret asked Einstein if he would like to earn £26 for giving the Deneke Lecture (in commemoration of Philip Maurice Deneke, Margaret's father)—but stipulated that the lecture must contain no math. The day would never come, answered Einstein, when he would *want* to lecture. Giving a talk was bearable, but later on, having to stick to what was said and written down, that was the awful part. (But he did agree to give the lecture.)

Margaret then showed Einstein a violin that someone wished to sell for £40. Einstein eagerly tried it out, but he advised her against buying it because its tone was very uneven. "He advised me to buy a cheap fiddle for 25 Marks instead and to give it to the man who had fixed up a fiddle for him, adding varnish, cutting away little bits, etc. and improving its tone. He had no other violin; but in California he had seen old Italian violins—powerful, and yet sensitive. The 'treated,' cheap fiddle was very thin in its wood, with a clear voice—not, of course, up to an Italian fiddle, and it had cost 200 Marks for [the] treatment and 20 for the fiddle."[3]

The lively and wide-ranging conversations continued during dinner. Afterward, the four string players played several Mozart and Haydn quartets, before Margaret joined them to play the Brahms Piano Quintet, opus 34. This is not an easy piece, and Einstein grumbled about having too many rests to count and faking so much that he could become a banker. It was 11:20 p.m. by the time they finished playing, and Einstein suddenly realized that the college gate was about to be locked. Margaret quickly telephoned the Christ Church porter and persuaded him to keep the gate open a while longer, while Helena got the car out to drive Einstein home.

*   *   *

A few weeks earlier, while he was still at Caltech, Einstein had had an interesting chat with the educator Abraham Flexner. Though he did not mention this conversation in his diary, it proved to be fateful. A philanthropic foundation had commissioned Flexner to create a new research institute that would allow scholars from different disciplines to work without being distracted by teaching or administrative duties.[4] Flexner—who had met Einstein briefly in New York—had come to Caltech to discuss this project with Millikan, who referred him to Einstein as someone interested in bold new approaches to education. Flexner called on Einstein in the Athenaeum, and, following a lengthy dialogue, they agreed to meet again in April, when Einstein would be in Oxford.

When Flexner arrived in Oxford, he went to see Einstein at Christ Church, and the two men chatted as they strolled on the college meadow. At the end of their conversation, Flexner offered Einstein a post at the nascent Institute for Advanced Study in Princeton. Einstein asked for time to think it over, for he was not enthusiastic about leaving Europe.

Einstein left Oxford and returned to Germany in June, spending the remainder of the summer at his Caputh retreat. That is where Flexner found him later that summer. With the Weimar Republic now in its death throes, Einstein accepted Flexner's offer, promising to take up his new position at the institute in the fall of 1933.[5] He had, accordingly, already made provisions to emigrate to America when the National Socialist revolution took place in January 1933.

In preparation for his third trip to Pasadena in December, Einstein applied for a visitor's visa at the US Consulate in Berlin—something that his travel agents had attended to on previous occasions. The visa was issued, but not before Einstein had been interrogated closely regarding his dubious political associations and his pacifist principles, and only after he had walked out of the consular office in a huff.[6]

AT SEA AGAIN

Phillip Frank has related an incident that took place before the start of the Einsteins' last transatlantic voyage, as they were locking up their home in Caputh. Einstein told Elsa to take a good look at their villa, for she would never see it again. Elsa thought that his premonition was foolish.[7]

In planning his third trip to Pasadena, Einstein again chose the long sea route: he and Elsa were to sail from Bremerhaven directly to Los Angeles by way of the Panama Canal—steering clear of New York. Einstein's friend and physician Janos Plesch, a wealthy celebrity doctor in Berlin, picked the pair up in Caputh and drove them to the Lehrter Bahnhof, Berlin's main railroad station. A group of friends had gathered there to see the pair off, and after saying their farewells, Einstein and Elsa boarded the train to Bremen. As in the year before, the Hapag agent Herr Geissler, Einstein's old schoolmate, met them at the railroad station and escorted them to the dock where the MS *Oakland* was tied up. The ship was only three years old and of similar size (6,500 tons) as the *Portland* of the year before.[8] The Einsteins brought several boxes of books on board, nearly all of a nonscientific, literary nature, and mostly recently published volumes—according to an inquisitive *New York Times* reporter.[9]

Einstein began what was to be the last of his travel diaries on the day of their departure (December 10, 1932), but he maintained it for only a week.

Once aboard, Einstein and Elsa were shown to their very attractive cabins on the upper deck, and they were pleasantly surprised to discover that they knew both the ship's doctor and their steward from a previous passage. The weather was cold and calm when the *Oakland* left the dock at 4:00 p.m. and steamed down the Weser River into the North Sea. Heading in a westerly direction, she followed the Dutch coast and reached the lock leading to Antwerp's inner harbor late the following evening.

In the morning (December 12), Einstein's uncle and aunt, Cesar and Suzanne Koch, who lived in nearby Liége, came on board. Einstein and Elsa were delighted to see these congenial relatives again and to exchange news and reminisces with them. The Kochs were not Einstein's only visitors, for in the evening, the theoretical physicist Paul Langevin appeared. Einstein had known and admired the 'splendid' Langevin for a long time, both as a fellow scientist and as a like-minded pacifist. On this occasion, Langevin came to confer with Einstein on how the community of prominent intellectuals in all European countries could promote international goals effectively.[10] By the following afternoon, all visitors had left the ship, and Einstein was alone on deck, observing the hustle and bustle in Antwerp's harbor. It seemed to him that it was significantly busier than the year before.

The *Oakland* served as a freighter as much as a passenger liner, and while her cargo was being loaded, Einstein turned to his travel library. He read a newly published pamphlet by Franz Oppenheimer, the economic sociologist, who was a personal friend. The pamphlet, *Neither Thus, Nor Thus*, rejected the notion that suffering humanity had to choose between personal freedom with the attendant inequalities on the one hand and equality, with personal freedoms crushed, on the other. His pamphlet offered a "Third Way" attainable only through unity.[11] It affected Einstein deeply, for the writing was remarkably clear and unaffected, and it contained a great deal of truth.

Physics was next on Einstein's agenda. He focused his attention on particle waves; specifically, he attempted to replace Dirac's quantum-relativistic equations with simpler ones. He was bothered by the asymmetry of Dirac's spinors, as well as by the empirical introduction of the mass constant—although that seemed unavoidable to him.[12]

By December 14, the *Oakland* had finished loading cargo, and she left the dock in wet and foggy weather. After she passed through the new lock into the Schelde River, the fog became so thick that she was obliged to spend the night at anchor in the river. By morning, the fog had lifted, and the *Oak-*

land continued down the Schelde into the English Channel, passing the Isle of Wight the following day. Einstein was delighted to discover that the ship's captain came from Baden, the same region of Germany that he and Elsa hailed from, and that he was, moreover, a reasonable and liberal-minded person. Einstein and the captain would have several interesting conversations. The *Oakland* was carrying more cargo, mostly steel rods, than the *Portland* the year before, and also she carried a few more first-class passengers (fourteen). All this suggested to Einstein that the economic crisis might be abating. But the news from Germany continued to be discouraging: Kurt von Schleicher had succeeded von Papen as chancellor and had come out in favor of rearmament—'always the same rascals,' commented Einstein.[13]

Surprisingly, this was the only political comment Einstein made in his diary during these final, turbulent weeks of the Weimar Republic. Ever since last July's Reichstag election, Einstein seemed to have taken it for granted that Hitler and the Nazis would come to power. He was soon to find himself in the ambiguous position of being celebrated in California as the great German scientist while he was vilified at home as a Jew. To add to this incongruity, Einstein's sojourn in Pasadena was funded by a foundation whose mission was the promotion of amicable German–American relations![14]

Both Einstein and Elsa suffered from stomachaches aboard the *Oakland*, even though they had only high praise for the ship's cuisine. Einstein found solace in reading the 'beautiful fairy tales' of Hans Christian Andersen, even though, in Einstein's view, his stories lacked the grand concepts that are found in true folk legends. Andersen dealt mostly in somewhat dusty trivia, according to Einstein, and the 'sentimental glorification of poverty,' which haunted the nineteenth century, plays an important role in his stories. Nevertheless, Anderson was an inspired poet, asserted Einstein, and the story of the Chinese nightingale ("The Nightingale") was his favorite.

After four days at sea, the weather cleared. Einstein studied the just-published book on abstract algebra by the influential Dutch mathematician Van der Waerden, and he was impressed by its ingenuity.[15] Max Born's treatise on quantum mechanics was also on his reading list, as it had been on his last voyage, but the statistical interpretation of quantum mechanics it presents—long a bone of contention between Bohr and him—continued to strike Einstein as 'artificial.'

PASADENA ONCE MORE

The preceding comments were made on December 18, in what was to be Einstein's last entry in his travel diary for six weeks. When he took up the diary again, he and Elsa had settled in Pasadena and he was already immersed in the academic, social, and musical life that he usually led there. It is therefore necessary to turn to sources other than his diary, such as contemporary newspaper accounts, in order to keep track of his activities during the six-week interim.

After the *Oakland* dropped anchor in Los Angeles harbor on January 9, 1933, Einstein, Elsa, and their thirty pieces of luggage passed through quarantine and customs quickly, thanks to the efforts of the Hamburg America Line's district manager. The welcoming delegation that awaited them on shore included Einstein's colleague and interpreter Richard Tolman, as well as Robert Millikan, Arthur Fleming, the local German consul, and several others. At the press conference that followed, Einstein lent his support to Tolman's theory of a pulsating universe, which envisions the expansion of the universe to continue until gravity causes it to contract and, eventually, to collapse, giving rise to the next big bang (see chapter 5). Asked about the nature of cosmic rays, Einstein deferred to Millikan, who then boldly predicted that his ongoing experiments would reveal the nature and origin of cosmic rays within a year. Einstein parried all questions regarding his own research work but enlarged on an idea for calming the struggle between capital and labor by curtailing the production of machinery, which he held responsible for the rising unemployment. In his view, the overuse of powered machinery led to a surplus of labor and was the main cause of the current economic distress—a position Oppenheimer had taken in *Neither Thus, Nor Thus*, and one reminiscent of the Luddites.

After the press conference, Einstein and Elsa were driven to the by-now familiar Athenaeum, where luxurious accommodation had been prepared for them. The suite, their home for the next two months, consisted of a dining room, a living room, and two bedrooms. It also included a sunny porch where Einstein was observed to sit for several hours the next day, smoking his pipe and reading a book. Elsa, meanwhile, went shopping with her friend Lili Petschnikoff, and when she returned, she presented her husband with a dozen California oranges.

On the day he arrived in Los Angeles, Einstein had agreed to take part in a huge meeting organized by the students of all Southern California univer-sities. This symposium, entitled "America and the World Situation," was meant to address the worldwide economic depression. The January 22 event drew an audience of several thousand people to Pasadena's Civic Auditorium and was broadcast nationwide by the National Broadcasting Company. The principal speakers, apart from Einstein, were the banker and economist Henry Robinson and the historian Henry Munro. The economic remedies offered by the three savants sound less than practical today: Robinson would end the worldwide depression by eliminating fear, while Munro saw the solution in the suspension of politics and an appeal to Congress to acquire just half as much good common sense as the suffering citizenry had demon-strated. Einstein, finally, blamed the depression on the rising utilization of new machinery: it made industry increasingly efficient, and it caused unem-ployment to rise and purchasing power to decline.[16]

With this, his most public function, behind him, Einstein resumed making almost daily diary entries, offering a glimpse into his daily routine in Pasadena.

On January 28, Einstein calculated certain solutions of the electromag-netic field equation in terms of spinors (vectors with complex components, which are useful in theoretical physics, particularly in dealing with particles with intrinsic angular momentum—spin). On Sunday evening, he and Elsa visited Lili Petschnikoff in her Hollywood home, where Einstein played music. On Monday he lunched with Judge Julian Mack, a prominent advo-cate of Jewish causes, and discussed various Zionist issues with him, before joining Upton Sinclair in the evening to hear a young journalist's lecture about contemporary Russia.

In the morning, Einstein paid a visit to the astronomer Charles St. John, who was suffering from heart disease but had managed to 'keep his free spirit.' In the afternoon, Einstein gave his second lecture, on "semi-vectors," which he considered 'more natural' than the spinors used by Dirac and Pauli. The violinist Fritz Kreisler happened to be in Los Angeles, and at his concert that evening, Einstein was particularly impressed by his rendition of Tartini's "Devil's Trill Sonata." At the astronomy seminar on Wednesday morning, Einstein heard a talk about the newest, most precise ruling engine for fabri-cating diffraction gratings. Gratings are key components of high-resolution spectrographs sought after by physicists and astronomers alike: the more lines per inch, the greater the grating's spectral resolution. Einstein spent the afternoon and evening quietly in his room. He grumbled that Gustav Struve, the German consul, was pestering him regarding his participation in a banquet to promote international understanding.

Einstein and Elsa spent the following weekend in the desert, this time as guests of the Hirsch family, who picked them up and drove them to their estate in Palm Springs. There they were introduced to Alfred Hirsch, the former conductor of the San Francisco Symphony, 'a cheerful and kind-hearted man,' according to Einstein. He and Elsa joined the family for an automobile excursion on Sunday, driving into the mountains and to an ancient stone monument featuring pictographs made by early indigenous Americans, most likely the Andreas Canyon site. They were also taken to see a lake, whose name Einstein could not recall, but which was almost certainly the Salton Sea, a large saline rift lake—an inland sea in the middle of the desert. Einstein thoroughly enjoyed the excursion and the wonderful desert weather. Early on Monday morning (February 6), Alfred Hirsch drove Einstein and Elsa back to Pasadena.

That same day, the 'young theoretical physicist' Robert Oppenheimer came to chat with Einstein in the Athenaeum. If Einstein recognized Oppenheimer's brilliance, he made no mention of it in his diary. Einstein then heard a lecture by the physical chemist Linus Pauling, who was among the first to employ quantum mechanics in the study of molecules; on this occasion, Pauling presented the electronic structure of the benzene ring.[17] In the evening, Einstein and Elsa dined again at the home of Upton Sinclair, to whom they appear to have become quite attached.

On Tuesday, Einstein worked with Tolman and visited the laboratory of Rudolph Langer, where he saw the latest experimental equipment (cloud chambers and magnets) for observing cosmic ray particles and determining their mass and charge. In the evening, Einstein and Elsa dined again at Charlie Chaplin's home, where Einstein and two other musicians joined Chaplin in playing Mozart string quartets. (This may have been the only occasion on which Einstein played with a left-handed violinist.) Einstein observed that one of the guests was a fat lady who befriended celebrities, professionally.

Einstein devoted Wednesday morning to catching up on his voluminous correspondence, after which he met with Caltech's Thomas Morgan, the most important American geneticist at the time. The two men discussed the organization of a scientific institute and the role played by foundations—

probably with the nascent Institute for Advanced Studies in Princeton in mind. Einstein also met briefly with Theodore von Kármán, a pioneer in aerodynamical theory and another member of Caltech's distinguished faculty, as was Linus Pauling.[18]

In the evening, Einstein and Elsa headed again toward Upton Sinclair's house, where they spent the evening discussing the current economic situation and much else. Because of Einstein's growing interest in economic theory, he read Irving Fisher's latest book on debt deflation, and he went to meet the author the very next day.[19] In his book, Fisher blamed the current economic crisis on excessive debt, but since his analysis went no further than that, Einstein asked where the excessive debt came from.

Getting back to the world of physics on Friday, Einstein heard the theoretical physicist Ralph Fowler, who was visiting from Cambridge, talk about the ongoing research in nuclear physics in Ernest Rutherford's laboratory.[20] Einstein gave an informal seminar to Epstein, Tolman, and a few others the next day and presented his latest research effort: a theory of terrestrial magnetism—which, on closer examination, turned out to be untenable. He spent the afternoon and evening with a certain Dr. Hoffman from the University of California, who insisted that research funds could be obtained only by soliciting wealthy patrons—and then tried to enlist Einstein's service to that end.

Einstein's next week followed much the same pattern, offering more evidence of his wide-ranging intellectual curiosity. On Sunday, February 12, he and Elsa heard a concert of Wagner's music, with Bruno Walter as the guest conductor of the Los Angeles Philharmonic Orchestra. Afterward there was a benefit dinner for Jewish workers in Palestine with several speeches that Einstein, for once, thought were excellent. The event ended with Einstein signing photographs to be sold 'at retail' for $70 each.

In the same week, Einstein also heard Richard von Kühlmann, the former German secretary of state, give a talk about Germany's foreign policy. His account was, in Einstein's view, clever and objective—a little too objective—and the speaker, too detached. Afterward, the two men had an interesting conversation over lunch. At the weekly theoretical physics seminar, Einstein heard a review of Eddington's most recent theories, and, in the evening, he attended a dinner meeting of the Criminal Society, at which true and false twins (identical and fraternal?) were under discussion. The next morning, Einstein wrote up the results of his own work and sent them to Mayer. In the afternoon, he conferred with a group of pacifists about the current political situation. His day ended with dinner at someone's home, followed by chamber music, which Einstein played on a marvelous borrowed violin. He made another trip to the Mount Wilson Observatory the following day. There he was shown Horace Babcock's latest ruling engine for fabricating the high-resolution diffraction gratings used to record stellar (and solar) spectra.

Einstein reported these and similar activities, albeit only very briefly, in his travel diary until February 16, when he put it aside for good. Surprisingly, the dramatic events taking place in Germany at that time received no mention: on January 30, 1933, Hitler was appointed chancellor and quickly gained complete control of the government. The National Socialist revolution had begun, and with it, the brutal persecution of political foes and Jews.

Einstein's remaining four-week sojourn in Pasadena followed its usual course. He and Elsa were again guests of Samuel Untermeyer at his Palm Springs estate over the weekend of February 26, and, as always, Einstein relished the desert sunshine enormously. Untermeyer offered to take him and Elsa back to Los Angeles in a plane on Monday morning, but they opted to remain on solid ground and returned to Pasadena by automobile.

On March 2, Einstein spent the afternoon at the Hollywood home of the sculptor Frederick Schweigardt, who modeled Einstein's likeness in clay, later to be cast in bronze. Schweigardt, a student of Auguste Rodin, had emigrated to America in 1930 and was a native of Lorch, a town in the part of Swabia from which Einstein and Elsa's family came. According to a newspaper report, Elsa "chatted volubly" throughout the afternoon with both the artist and her husband while the work was going on.[21]

BACK TO EUROPE

On Friday, March 10, one day before Einstein left Pasadena, he held a news conference in the Athenaeum at which he announced his decision not to go home to Berlin. From now on, he was willing to reside only in countries that preserved "political freedom, tolerance, and equality before the law of all citizens." He confirmed that he planned to be in Princeton from October to March 1934 and that he would be affiliated with a great new institute for theoretical investigations. Richard Tolman was seated on Einstein's right to help him frame his answers in English, but Einstein's English had improved to such an extent that his help was scarcely needed.

On Saturday evening, Einstein and Elsa boarded the Santa Fe Railroad's California Limited for Chicago. Elsa was attired in a new, fur-trimmed coat and wore a corsage of lilies-of-the-valley, while Einstein clutched his violin as the pair said good-bye to the many well-wishers who had come to the station to see them off. Among them were Hubble, Tolman, von Kármán, and Epstein with their families, as well as the consul, Gustav Struve, who presented Einstein with some sort of diplomatic travel document.[22]

Einstein and Elsa arrived in Chicago on Tuesday morning and were immediately whisked off to the city's exclusive Standard Club, where Einstein conferred with a delegation of peace advocates headed by the famed labor and criminal lawyer Clarence Darrow. At noon there was a public benefit luncheon to celebrate Einstein's fifty-fourth birthday and to raise money for the Hebrew University in Jerusalem. This dazzling affair was attended by the cream of Chicago's dignitaries and was presided over by the physicist Arthur Compton. By 3:00 p.m., the guest of honor and his wife were again back on a train, now heading for New York, where they arrived the following day. Paul Schwarz, the German consul known to the Einsteins from their previous visit to New York, boarded their train in Albany and escorted them to Grand Central Station where, upon Einstein's earnest request, *no* official reception took place.

From the station, Einstein and Elsa were driven to the Hotel Commodore, where that evening Einstein's birthday was celebrated once again. A thousand guests attended this celebration and helped to raise funds for the Hebrew University and for the Jewish Telegraphic Agency, an international news service whose chairman was Jacob Landau. At the dinner, Samuel Untermeyer, the Einsteins' Palm Springs host, read a poem he had composed for the occasion, and Einstein gave a speech that avoided all politics and demonstrated what a master of the cliché he had become. The *New York Times* printed its text in full, as well as the speeches of the other principal speakers: Karl Compton (Arthur's brother), president of the Massachusetts Institute of Technology, and the astronomer Harlow Shapley, who spoke of the past few years' spectacular discoveries regarding our universe.

Two days later (March 17), at a reception hosted by a pacifist group, Einstein appealed for "moral intervention" against Hitlerism—because he feared that a direct anti-German intercession would be harmful. Amid this whirl of activities, he also found time to act as godfather to Mr. Landau's newborn son, at the boy's circumcision (bris)—which the *New York Times* coyly referred to as a "ceremony of initiation held on the eighth day after birth, according to Jewish custom." Afterward, Einstein and Elsa traveled to Princeton, where they had lunch with Oswald Veblen, who would soon be Einstein's colleague at the Institute for Advanced Study. The Einsteins also met with Luther Eisenhart, the dean of Princeton University, who showed Einstein the temporary office in Fine Hall that he would occupy in the fall. Eisenhart and his wife then drove the couple around Princeton to point out to them the different residential areas that the town afforded.

Instead of returning to Germany on the Hapag liner *Deutschland*, as originally planned, Einstein and Elsa booked passage to Antwerp for March 18 on the Red Star liner SS *Belgenland*—the ship that had brought them to Pasadena the first time, in 1930. A group of about a hundred women ranging from fifteen to seventy-five years of age, who represented several peace organizations, were waiting for Einstein on board the *Belgenland* and applauded when he arrived. He agreed to meet with them in the ship's lounge after he was settled in his cabin. On his return, an impromptu symposium ensued in which he answered many questions and also made some remarkable pronouncements: that an unarmed country would never be attacked by another country, and that the possibility of a conflict between the United States and Japan should not be taken seriously. He then locked himself in his cabin and emerged only after all visitors had gone ashore. As the *Belgenland* pulled away from the pier, he was back on deck, smiling and waving his arms to the plaudits of the women ashore.

After a brief stop at Le Havre, the *Belgenland* arrived in Antwerp on March 28. The city's mayor came aboard to welcome Einstein and Elsa, and when the pair walked down the gangplank an immense crowd cheered, "Long live Einstein!" Asked about the recent events in Germany, Einstein blamed the current anti-Semitic outbreaks on emotion overcoming reason, and he repeated his resolve not to return to Germany for as long as "this pathological situation" persisted.

LAST DAYS IN EUROPE

In the ten days it took Einstein and Elsa to return aboard the *Belgenland* (March 18–28, 1933), events with far-reaching consequences had taken place in Germany. Following the torching of Reichstag building, the newly appointed propaganda minister Joseph Goebbels organized a spectacular military ceremony in Potsdam, the citadel of the Hohenzollern dynasty. On March 21, dubbed the Day of Potsdam, a series of ceremonies designed to portray Hitler's regime as the natural successor to the monarchy took place. While standing next to the eighty-six-year-old Hindenburg, Hitler convened the Reichstag, with great pomp, in the very Potsdam church in which Frederick the Great is buried. With the needed support of the conservative parties, the Reichstag passed the empowerment statute (*Ermächtigungsgesetz*) two days later, the only opposition coming from the ninety-four Social Democrat delegates.[23] The empowerment statute stripped the Reichstag of its legislative power and conferred dictatorial powers on the government—in reality on the person of Hitler. The day's festive rites concluded with a command performance of *Die Meistersinger* at the state opera.

Finding himself now—at journey's end—a voluntary exile from the country of his birth, Einstein quickly came to grips with the new situation. On March 28, the day the *Belgenland* arrived in Antwerp, Einstein hired a taxi to take him from the harbor to Brussels, where he called on the German consulate. There he handed in his German passport and renounced his Prussian citizenship. The same day, he wrote his letter of resignation to the Prussian Academy, expressing his gratitude for nineteen years of intellectual stimulation and beautiful personal relationships with members of the academy, and informing the academy that it was intolerable for him to serve the Prussian government under the prevailing circumstances. When the letter arrived two days later, the government officials in charge of the academy were furious, according to Max von Laue, because it foiled their efforts to take action against Einstein. It is, indeed, a measure of how intensely the Nazis despised Einstein that just days after gaining power, the government instructed the Prussian Academy to prepare formal disciplinary proceedings against him![24]

Einstein was now unemployed, but although his home had been looted, and his bank accounts and property confiscated, his situation was hardly desperate. He had landed in Belgium, where he had many friends, and it was not long before professorships were offered to him in Oxford, Paris, and Madrid. It had long been his custom to keep his foreign earnings in banks abroad, so that he was now able to rent a modest vacation home in the Belgian seaside resort of Le Coq sur Mer. In that house nestled among the dunes, Einstein and Elsa were soon joined by their two indispensable companions, Walther Mayer and Helen Dukas.

Einstein postponed his planned trip to England for a few days in order to visit his younger son, Eduard, who had been admitted to the Burghölzli psychiatric clinic after his condition had worsened. It was to be the last time Einstein saw him. From Zurich Einstein traveled directly to Oxford, where he arrived on June 1, 1933. He conferred with Rutherford and with Lindemann, who had recently toured Germany to assess the calamitous plight of Jewish scholars under the Nazi government and to recruit several of the best available physicists for the Clarendon Laboratory. Einstein resided in Christ Church, as before, and that is where he prepared his prestigious Herbert Spencer Lecture, which he delivered in English. He also traveled to Glasgow to give a lecture.

Soon after returning to Le Coq, Einstein was again embroiled in political controversy. After years of fervently opposing militarism, he rejected that philosophy in face of the dire threat posed by Hitler, and he contended that an ideology was tenable only if it justified its consequences. In an open letter to a Belgian pacifist, published in the newspaper *Patrie Humaine*, Einstein abandoned his uncompromising opposition to war and stated that if he were a Belgian he would accept military service as his contribution to the preservation of European civilization. Needless to say, Einstein continued to be criticized for both his inconsistency and his failure to anticipate the present contingency.

Reports surfaced that a secret Nazi society had offered a bounty of £1,000 for "silencing" Einstein. Einstein himself was unconcerned, and he continued to take daily early morning walks across the dunes, but because assassinations by Nazi agents outside Germany were not uncommon at that time, the threat to his life was taken seriously. During his stay at Le Coq sur Mer, Belgian police detectives guarded him round the clock.[25] Ultimately, Einstein agreed to move to England, to the home of Commander Oliver Locker-Lampson near Cromer on the Norfolk coast, which was much safer than Le Coq sur Mer. Locker-Lampson, Einstein's host, was a member of Parliament who introduced a bill in the Commons to extend citizenship of the British Empire to all Jews living in Germany, but the bill failed to pass. Locker-Lampson also brought Einstein to meet Winston Churchill at his home in Chartwell one day. Over lunch in the garden, the three men discussed the menace posed by Hitler and the accelerating rearmament of Germany.[26] Churchill's political fortunes were at low ebb at the time, and no one could have foreseen the historic role he would play six years later.

Einstein's last public appearance in Europe took place on October 3, when he spoke in London's Royal Albert Hall at a huge rally to raise funds for the growing number of German Jewish refugees. Speaking in heavily accented English, he warned his audience of the dangers ahead while giving high praise to Britons for remaining true to their traditions of justice, democracy, and tolerance.

A few days later (October 7), Einstein was driven in utmost secrecy to Southampton, where a harbor craft brought him to the Red Star liner *Westernland* (sixteen thousand tons) bound for New York. Once on board he was reunited with Elsa, Dukas, and Mayer, who had boarded the ship in Antwerp. In deference to his newly acquired status as a refugee, Einstein made this transatlantic crossing in tourist class. He would not see Europe again.

Einstein had always insisted very emphatically that he was coming to Princeton in order to work and to teach, and that he wished to avoid all publicity and interviews. On October 17, when his ship arrived in New York harbor, Abraham Flexner, now the director of the Institute for Advanced Study, and two of the institute's trustees came aboard the *Westernland* during her quarantine stop and welcomed Einstein and his entourage to America. Einstein, carrying his violin, and the others were then quietly transferred to a special tugboat that brought them to the Battery, where an automobile was waiting to whisk the party off to Princeton. By the time the official reception committee, waiting at the 23rd Street Pier, became aware of Einstein's covert landing, he was well on his way to New Jersey.

In Princeton the four new arrivals took up temporary residence in a small hotel, the Peacock Inn. Later that afternoon, Einstein was observed taking a stroll while he smoked his pipe and chatted with Mayer at his side. He walked to pick up the evening paper at the corner newsstand, and afterward he bought himself a cone at a nearby ice cream parlor.[27]

Einstein had found a new home.

Epilogue (1933–1955)

Einstein's move from Berlin to Princeton represented, in many respects, a sharp discontinuity in his life. His image as a world-traveling public celebrity changed almost overnight into that of a somewhat eccentric professor cloistered in a New Jersey college town. But from Einstein's perspective, little of significance had changed: after making only minor adjustments, he continued to pursue his principal passions of physics, music, and social justice, while his craving for privacy became much easier to satisfy. Before long, he was even able to gratify his love of sailing when he bought a seventeen-foot sailboat, which he named *Tinnef* (Yiddish for "piece of junk") in wistful contrast to the magnificent *Tümmler* that he had left behind in Germany.

SETTLED IN PRINCETON[1]

When Einstein arrived in Princeton and was shown his temporary office in the university's Fine Hall, he commented that all he really needed for his contentment was a table and a chair, paper and pencil, and a large wastepaper basket to hold his many mistakes. In Fine Hall he was soon surrounded by new colleagues, as well as a number of familiar faces from the Old World, including Oswald Veblen, John von Neumann, Hermann Weyl, Eugene Wigner, and Rudolf Ladenburg. In 1934, Schrödinger also came to Princeton, but Einstein's efforts to have him join the institute's staff were in vain, largely because Einstein had by then come into conflict with Flexner, the institute's first director, and had even threatened to resign his post. The two men had clashed over Flexner's inflexible policy of shielding Einstein from all publicity and preventing him from participating in any public function. Flexner was fearful that such public attention might engender hostility toward German refugees and bolster anti-Semitic sentiment. He went so far as to open Einstein's mail and to intercept an invitation to Einstein from President Roosevelt to visit the White House. Fortunately, this particular outrage was subsequently rectified, and in January 1934, Einstein and Elsa dined with Franklin and Eleanor Roosevelt and spent the night in the Franklin room of the White House.[2]

In his domestic life, also, Einstein was once again watched over by the two women he had long been closest to. From the Peacock Inn, Einstein, Elsa, and the indispensible Helen Dukas moved to a rented house, and Einstein found his new hometown so much to his liking that he severed almost all his links to Europe. He wrote Lindemann that he did not wish to retain his Oxford research studentship, and he expressed the hope that another scientist could receive it in his stead. He was done with Europe.

When in the spring of 1934 word arrived that Elsa's daughter Ilse was seriously ill, Einstein was content to let Elsa travel to Europe by herself. He saw her off in New York, where she boarded the *Belgenland*, the liner on which she and Einstein had traveled twice before. Ilse was only thirty-seven when she died later that summer, and Elsa returned grief-stricken to Princeton. Einstein meanwhile spent the summer months in a cottage he had rented in Watch Hill, Rhode Island, together with an old friend, Gustav Bucky, who had come to America in the 1920s. Einstein enjoyed the informality of Watch Hill and took pleasure in sailing aimlessly, and alone, in the many coves of Narragansett Bay. He did not return to Princeton until October.

Having decided to make Princeton his home, Einstein was anxious to become an American citizen. This necessitated changing his visitor's visa to an immigration visa, a procedure that could be performed only outside the United States. Einstein recognized this requirement as an opportunity for another sea voyage: instead of traveling by land to Canada, he chose to travel by sea to Bermuda. On this expedition he was accompanied by Elsa; his stepdaughter, Margot; her husband, Dimitri Marianoff; and Dukas. They sailed on the *Queen of Bermuda*, a luxurious liner of 22,500 tons, only two years old, that made the passage to Bermuda in just forty hours. Einstein and his party arrived in Hamilton on May 27, 1935, and they immediately visited the US consulate to take care of their official business. Afterward they reconnoitered several large hotels that Einstein judged too public and pretentious before ending up at the modest Roseacre guesthouse. When journalists later questioned the proprietor, he at first denied all knowledge of Einstein before admitting that they were on the right scent. But in the end, Einstein's wish to shun all publicity was respected.[3]

In a restaurant, Einstein chatted with a German cook, who invited him to go sailing with him in his small boat; when Einstein had not returned after seven hours, Elsa became concerned and eventually found him at the home of the cook, who had prepared several German dishes for her Albertle.[4] Five days later, the whole party returned to New York on the *Queen of Bermuda*.

Three months after their Bermuda trip, Einstein purchased a conveniently located, modest clapboard house at 112 Mercer Street. Elsa went to work to make some necessary renovations, in particular to have an upstairs study for Einstein constructed—as she had done at Haberlandstrasse 5. On the walls of the study hung the same pictures of Newton, Faraday, and Maxwell as in Berlin, and here Einstein felt most at ease and spent most of his time for the next twenty years. With Margot's help, several pieces of furniture, including the grand piano, had been saved from their Berlin apartment, and these were used to furnish the rooms of their new home.

That summer (1935), Einstein and Elsa escaped Princeton's hot and humid weather by renting a vacation home in Old Lyme, Connecticut, where he spent many hours sailing in the estuary of the Connecticut River. They summered on Lake Clear in the Adirondacks the following year, but after returning to Princeton, the heart and kidney problems from which Elsa had suffered for some time grew more serious. Her death in December 1936 devastated Einstein. He missed her terribly.

Helen Dukas took over running the Mercer Street household, which included Einstein's stepdaughter Margot, a gifted sculptress, who had joined the establishment following her divorce, and, after 1939, Einstein's sister, Maja (Maria Winteler), who had been obliged to leave her home in Italy when Mussolini restricted the residency of foreign Jews. Maja's husband, Paul Winteler, meanwhile, returned to Switzerland. Einstein loved Maja dearly and was delighted when she arrived at 112 Mercer Street. He also persuaded and assisted his son Hans Albert to immigrate to the United States; after a reunion with his father, he and his family settled in Berkeley. Of Einstein's immediate family, only Mileva and his unfortunate son Eduard remained in Europe: Mileva died in Zurich in 1948, while Eduard lived on in the Burghölzli clinic until 1965.

PHYSICS, THE BOMB, AND POLITICS

As in Einstein's personal life, there was continuity in his scientific work. He pursued his controversy with Bohr by designing thought experiments to challenge the uncertainty principle and questioning the completeness of quantum mechanics. Unwilling to abandon the intuitive view of reality—either a particle is at a certain location or it is not—he did not accept the probabilistic quantum mechanical view, even though it was in accord with numerous experiments. It is noteworthy that in his youth, Einstein had done the opposite: he had rejected the commonsense and universally accepted view of space, time, and matter and had forged a new view of nature.

In 1935 Einstein, in collaboration with two young associates at the institute, published a celebrated article demonstrating the incompleteness of quantum mechanics.[5] Like his other barbs directed at quantum mechanics, it missed its mark; it did, however, stimulate widespread discussion among theoreticians and obliged them to pay closer attention to the fundamentals of quantum mechanics—which remain a subject of debate.

Einstein's primary preoccupation remained his search for a unified field theory, from which electromagnetism and gravitation could be derived. This problem occupied him for over thirty years but was, in hindsight, ill-conceived. To be sure, his quest made sense when it began, but he pursued it long after other forces of nature had been discovered and the quantum-mechanical view was well established. Einstein's strategy of using mathematics as a guide to nature, which had served him so well in relativity theory, now failed him. But he stuck to it with the same stubborn persistence with which he had overcome the daunting difficulties he encountered in formulating general relativity—this time, alas, without success.

Soon after arriving in Princeton, Mayer, Einstein's longtime calculator, abandoned him to pursue his own research. There were many young physicists at the institute who were glad to work with Einstein, however. With them he completed two significant contributions to physics: He showed that gravitational waves—although they have escaped detection so far—exist as a consequence of relativity and that the Laws of Motion are contained in the gravitational field equation, not distinct from gravity, as is the case in Newtonian mechanics.[6] This work apart, Einstein withdrew largely into his own world of ideas and took little interest in the exciting developments in elementary particle physics. When Bohr visited the institute for several months in 1939, Einstein avoided him. Sadly, these two giants of twentieth-century physics had only a cursory, formal conversation with each other.[7]

As noted, Einstein's pacifist principles fell by the wayside as the world headed toward war again. In the summer of 1939, while vacationing at

Nassau Point, Long Island, he was visited by the two theoretical physicists, Eugene Wigner and Leo Szilard, who drew his attention to recent experiments indicating that neutrons could initiate a chain reaction in uranium with the release of enormous amounts of energy, according to Einstein's formula $E = mc^2$. That visit led to a series of discussions culminating in Einstein's famous letter to President Roosevelt, which alerted him to the military implications of these discoveries. It is unlikely that Einstein's letter did more than raise the government's awareness of nuclear weapons, but two years later the Manhattan Project came into being. Einstein later commented that had he known that the fears of Germany's building a nuclear bomb were groundless, he would not have written the letter. It pained him that some people held him culpable for the horror of Hiroshima.[8]

For a long time, Einstein was keenly aware of his status as a refugee in America, and he was circumspect in his support of public issues, but with the end of the war and the dropping of the bomb, he was ready to exercise his rights as an American—he had been a naturalized US citizen since 1940. He campaigned passionately for a world government with the military resources to prevent future wars. Many considered his position to be naïve, but Einstein defended it as realistic and as offering the best chance for avoiding nuclear war. At Szilard's urging, Einstein agreed to chair the Emergency Committee of Atomic Scientists, and he took part in countless discussions and proclamations on behalf of world federalism. But as the shadows cast by the cold war lengthened, his efforts ultimately had little tangible effect.

Having twice experienced horrific wars instigated by German militarism, Einstein was unforgiving toward his former 'step-fatherland' and opposed the reindustrialization of Germany. He rebuffed attempts by former colleagues in Germany to reinstate him in the academic institutions that had expelled him under the Nazis, and he maintained contact only with Max von Laue and a few other former German associates. After Nernst and Planck died, Einstein did, however, pay touching homage to these men who he had esteemed highly as colleagues and deemed to be good men, and he wistfully recalled the international spirit that had permeated science in better times.[9]

ENDGAME

Einstein resigned his post at the institute in 1944 when he reached his sixty-sixth birthday, but he maintained his routine of walking to his office every morning. His strolls along the streets of Princeton gave rise to the numerous stories—some of them true—that Princetonians tell about the kindly, absent-minded recluse who lived among them. He walked alone or in the company of Kurt Gödel (to whom he was very close), Pauli, or other German-speaking colleagues or friends, for while Einstein rejected all formal connections to Germany, he could not escape the strong bonds that tied him to the German language, culture, and music.

Even though his ability to affect the political climate waned after the war, the fervor with which he followed his humanitarian instincts went undiminished. He made a point of accepting an honorary degree from Lincoln University, a black institution, and spoke out forcefully against racism; and when Marian Anderson gave a concert in Princeton in 1937 and was refused a room at the Nassau Inn, he invited her to stay with him. In 1948, when the United States still enjoyed a monopoly on nuclear weapons, he commented in a letter to his old friend Toni Mendel that wherever he set up his tent, people seemed to become militarists and war-fanatics, and that it was now the turn of his American compatriots, who seemed to have inherited it from the Prussians.[10] Five years later, at the height of the McCarthy witch-hunts, which were based on the flimsiest evidence, Einstein was outspoken in his defense of civil liberties and warned of the dangers of the growing fascist mentality. When Toni applauded him for publicly taking that stand, he replied that her letter showed him that the fire in her belly was still burning fiercely and that he was reminded of those times long ago, when they had been on the warpath together.[11]

Comfortably ensconced in 112 Mercer Street, Einstein mellowed during his last years, but he remained the loner he had always been. As he put it: "I wished for this isolation all my life, and now I have finally achieved it here in Princeton."[12] He had many visitors, both old friends and celebrities, and was lovingly cared for by a trio of women, all utterly devoted to him: Helen Dukas; his sister, Maja; and his stepdaughter, Margot. Einstein's health was slowly deteriorating. In 1948, following several painful abdominal attacks, he was admitted to a hospital, where a grapefruit-sized aneurysm was discovered in his abdominal aorta. Maja's health was worse than his, and when she suffered a stroke and was bedridden for many months, Einstein read to her every evening. Among the books he chose was Cervantes's *Don Quixote*—with whom Einstein, admittedly, shared a penchant for tilting at windmills. Maja died in 1951, but Helen Dukas and Margot outlived Einstein.

On April 11, 1955, just two days after he signed Bertrand Russell's appeal for an end to the nuclear arms race, Einstein's aneurysm perforated. In great pain, he was hospitalized, but he resolutely rejected any attempt to prolong his life as being "tasteless." Five days later, Einstein died, "gracefully," as he had wished.

Notes

PREFACE

1. For the English text of the Einstein-Tagore conversations, including a commentary, see Wendy Singer, "'Endless dawns' of imagination," *The Kenyon Review* 23, no. 2 (Spring 2001): 7–33.

2. Andor Carius, *Das Goldene Boot*, ed. M. Kämpfchen (Düsseldorf: Artemis & Winkler, 2005): 648–55.

3. Letter in possession of Gerald Mendel.

ACKNOWLEDGMENTS

1. The Albert Einstein Archive Call Numbers of the travel diaries, and the chapters devoted to each, are as follows. 5-253.00: Chapters 2 and 3. 5-255.00: Chapter 4. 5-256.00: Chapter 5. 5-257.00: Chapter 6. 5-258.00: Chapter 7. 5-259.00, 260.00, and 29-144.00: Chapter 8.

1. SETTING THE STAGE

1. Among the other scientists Einstein met with in Berlin were Walther Nernst, an early champion of Einstein's work on quantum physics; Fritz Haber, a brilliant chemist and Nobel laureate; Erwin Freundlich, a young astrophysicist intent on testing the predictions of general relativity; and Max von Laue, the discoverer of X-ray diffraction, who was philosophically allied with Einstein. These men, along with Planck, were probably already eyeing Einstein with a view to luring him to Berlin.

2. For more detailed accounts of the events leading up to WWI, see Thomas Levenson, *Einstein in Berlin* (New York: Bantam Books, 2003), pp. 50–60. Also see Robert K. Massie, *Dreadnought: Britain, Germany, and the Coming of the Great War* (New York: Random House, 1991).

3. In 1914, Corporal Adolf Hitler participated in the failed attempt to execute the Schlieffen plan. He would meet with greater success in 1940.

4. Albert Einstein, *Mein Weltbild* (Zurich: Europa Verlag, 1934), p. 9.

5. The biographical sketch that follows relies on several well-documented Einstein biographies: Albrecht Fölsing, *Albert Einstein: A Biography*, trans. E. Osers (New York: Viking, 1997); Philipp Frank, *Einstein: His Life and Times* (New York: Knopf, 1947); and Walter Isaacson, *Einstein: His Life and Universe* (New York: Simon and Schuster, 2007).

6. Albert Einstein, "Autobiographisches," in *Albert Einstein, Philosopher-Scientist*, ed. P. A. Schilpp (New York: Tudor Publ. Co., 1949), p. 4.

It is interesting that Francis Crick, arguably the greatest biologist of the twentieth century, became a lifelong atheist at about the same age, based on very similar reasoning. He lost faith and refused to attend church because science had revealed to him that some of the assertions in the Bible were false, and if some of the Bible is wrong, why should the rest be accepted? M. Ridley, *Francis Crick: Discoverer of the Genetic Code* (New York: HarperCollins, 2006), p. 9.

7. Einstein, "Autobiographisches," p. 8.

8. Einstein's insistence on the need to work on the problems presented by quanta found little support. The following story reveals his feelings about quantum physics at the time. Einstein's office in Prague looked out over the garden of an insane asylum, where inmates could be seen strolling about. He would take visitors to the window and explain that down there were those lunatics who *do not* concern themselves with quantum theory. Fölsing, *Albert Einstein*, p. 283.

9. The brilliant physical chemist Fritz Haber (1868–1934) is best known as discoverer of the Haber-Bosch process, by which nitrogen from the air is "fixed" and ammonia is synthesized. He received the Nobel Prize for this work. The process is of great economic importance in the production of fertilizer and munitions. Haber made many other contributions to chemistry and played a prominent role in the German war effort during WWI, particularly by organizing the first deployment of poison gas. He was a close personal friend of Einstein's, even though their political views differed profoundly. Because of his Jewish roots, Haber resigned from his institute after Hitler came to power, even though, as a war veteran, he could have stayed on a little longer. He died in Switzerland a year later.

10. The putative physics institute was under the aegis of the Kaiser Wilhelm Society, which was dedicated to scientific research. Founded in 1911, it is among the lasting accomplishments of Wilhelm II. The institutes supported by the society—now known as the Max Planck Society—flourish to this day. The kaiser had learned from visiting Americans that wealthy, prominent businessmen such as Carnegie, Rockefeller, and Guggenheim were donating large amounts of money to foundations that supported artistic and scientific work, and he recognized the importance of research for advancing Germany's power. The society he lent his name to was funded by industrialists and bankers but was ultimately controlled by the govern-

ment. After paying an initiation fee of twenty thousand Marks, members of the society were awarded the title of "senator" and the right to wear a handsome academic gown. They were, occasionally, invited to join the kaiser at breakfast, where the need for additional research funding was explained to them. Breakfast with Wilhelm was not cheap. Frank, *Einstein,* p. 106.

The Institute for Physics, directed by Einstein, who had an aversion to administrative duties, came into existence—at least on paper—in 1917. It was housed in Einstein's attic study at Haberlandstrasse 5. Since it had neither a building nor a staff, it confined itself to funding deserving research projects, and one of its first grants went to Freundlich for experiments to test a prediction of general relativity.

11. Chaim Weizmann (1874–1952) was an organic chemist who made important contributions to the Allied war effort in WWI and played a key role in the genesis of the Balfour Declaration.

12. Thomas Levenson, *Einstein in Berlin* (New York: Bantam, 2003), p. 259. For an account of how Einstein's visit was treated by the press, see *Albert Meets America*, ed. József Illy (Baltimore: Johns Hopkins University Press, 2006).

13. *The New York Times*, April 3, 1921.

14. *The New York Times,* April 14, 1921.

15. In 1924, both Philipp Lenard (1862–1947) and Johannes Stark (1874–1957) declared their adherence to Hitler and the Nazi Party, and though they were initially influential, they fell into political disfavor after 1933. They espoused a "German physics," which eschewed theory and stressed intuitive observation. Lenard's four-volume *Deutsche Physik*, published in 1936, makes no mention of relativity or of quantum mechanics, which he considered "Jewish physics." See Fölsing, *Albert Einstein*, p. 464 and p. 523. A riveting account of Lenard's and Stark's careers and of other bizarre aspects of science, medicine, and mathematics under the Nazis is found in Walter Gratzer, *The Undergrowth of Science* (Oxford: Oxford University Press, 2000), pp. 219–80.

16. Harry Graf Kessler, *Tagebücher 1918–1937* (Berlin: Deutsche Buch-Gemeinschaft, 1967), pp. 278–80. (Translated here by the author.) Kessler lived from 1868 to 1937.

17. The negotiations mentioned by Kessler resulted in the Treaty of Rapallo, which obliged Germany and Soviet Russia to abandon all claims for war reparations. The treaty caused considerable concern among the Allies and in Poland and was regarded as treachery by German nationalists.

18. Among the Einsteins' other guests were Max Warburg (1867–1946), the president of the influential family bank M. M. Warburg in Hamburg, and an erstwhile confidant of Wilhelm II; the "super-rich" banker and philanthropist Leopold Koppel (1854–1933), who was a generous contributor to scientific research; Erich Mendelsohn (1887–1953), the architect of the modernist solar observatory known as

the Einstein Tower; and Bernhard Dernburg (1865–1937), a financier and member of the Reichstag who had served as Germany's colonial secretary before WWI.

19. Einstein asked Kessler to repeat Painlevé's message several times, for Painlevé claimed to have found a solution of the Einstein field equation that corresponds to what is now known as the "event horizon" of a black hole. Kessler reports that Einstein considered the matter of little importance. During his visit to Paris, Einstein did lecture at the Collège de France, but he canceled the lecture at the French Academy of Science after he was told that some of the Academicians planned to walk out to protest the presence of a German professor at the academy.

2. JOURNEY TO THE FAR EAST (1922)

1. Maximilian Harden was the pen name of Felix Ernst Witkowski (1861–1927), a prominent publisher who initiated the notorious Eulenburg affair. His revelations destroyed the reputation of Prince von Eulenburg (1847–1921), the closest friend and advisor of Wilhelm II, who was credited with having a moderating influence on the kaiser. The affair exposed homosexuality in the highest echelons of the army. Note that Harden, Erzberger, and Rathenau were all Jewish.

2. The first approach to Einstein was made by the theoretical physicist Atsushi Ishiwara on Yamamoto's behalf. Ishiwara had studied in Germany under Sommerfeld, Planck, and Einstein. He wrote to Einstein of Yamamoto's idea in August 1921, and Einstein received the formal contract for the lecture tour in January 1922. Throughout Einstein's stay in Japan, Ishiwara served as his guide, translator, and constant companion. See Hiroshi Ezawa, "Impacts of Einstein's Visit to Japan," *Association of Asia Pacific Physical Societies (AAPPS) Bulletin* 15 (April 2006): 3–16.

3. Michele Angelo Besso (1873–1955) and Lucien Chavan (1868–1942) were close to Einstein in their student days in Zurich and remained his lifelong friends. Besso, in particular, was also a valued colleague: Einstein's celebrated 1905 paper on special relativity is remarkable in that it contains *no* bibliographical references, and only a single acknowledgement, which expresses his indebtedness to "my friend and colleague M. Besso who steadfastly stood by me in my work . . . for many a valuable suggestion." Albert Einstein, "Zur Elektrodynamik bewegter Körper (About the Electrodynamics of Moving Bodies)," *Annalen der Physik* 17 (1905): 891–921.

4. Roger Highfield and Paul Carter, *The Private Lives of Albert Einstein* (London: Faber and Faber, 1993), p. 220.

5. The *Kitano Maru*, an 8,500-ton, single-funnel, mail steamship, was built in

1909 and was owned by the NYK shipping line. She was sunk by a sea mine off the Philippines in 1942.

6. On this, as on his other travels, Einstein did not leave his work behind. Soon after finishing work on general relativity, he began searching for a field theory from which both the gravitational and the electromagnetic fields could be derived—a unified field theory. It was a quest to which Einstein devoted the remainder of his life. He was not alone in this quest. The eminent German mathematicians Hermann Weyl (1885–1955) and Theodor Kaluza (1885–1954) also attempted (unsuccessfully) to generalize the geometry of four-dimensional space-time, using an approach that had succeeded in general relativity theory. In his diary, Einstein expressed the hope that it might be possible to cling to a field theory, even though he had doubts that *all* laws of nature can be expressed in terms of differential equations—presumably because he recognized the importance of quantum effects.

Einstein considered relativity theory to be more fundamental than Newtonian physics because Newton's mechanics was derivable from it. He also believed in the existence of an unending hierarchy of ever more fundamental theories. Abraham Pais, '*Subtle is the Lord . . .*' *The Science and the Life of Albert Einstein* (Oxford: Oxford University Press, 1982), p. 325.

It is worth recalling that in 1922, gravitational fields and electromagnetic fields were the only fields physicists knew, and both give rise to long-range forces with an inverse square dependence on distance ($1/r^2$). It therefore seemed reasonable to search for their parent theory.

Following Rutherford's 1919 discovery of the nucleus and later the discovery of nuclear forces, it became clear that these would need to be included in attempts to unify the forces in nature. Einstein showed little interest in the emerging nuclear physics, however.

For a more comprehensive, nontechnical discussion of Einstein's quest for unification, see Steven Weinberg, *Lake View: This World and the Universe* (Cambridge, MA: Harvard University Press, 2009), pp. 178–85.

7. The book was *Durée et Simultanéité à propos de la théorie d'Einstein* (Paris: Alcan, 1922). Earlier that year, Einstein and Henri Bergson (1859–1941), the influential French philosopher, had engaged in a public debate that some critics viewed as a confrontation between rationalism and intuition.

8. Einstein had written: "Everyone but you seems a stranger to me, as if they were separated from me by an invisible wall." Highfield and Carter, *Private Lives* (London: Faber and Faber, 1993), p. 87. The book Einstein read was Ernst Kretschmer, *Körperbau und Character* (Berlin: Springer, 1921).

9. Due to the Earth's rotation, the lateral speed of the surface is greatest near the equator. As a result, winds approaching the equator from north or south are deflected in an easterly direction by the Earth's rotation. This is the reason that cyclones (hur-

ricanes) rotate in a counterclockwise direction in the northern hemisphere and in a clockwise direction in the southern hemisphere.

10. Einstein's interest in the gyrocompass had its beginnings in 1914, when he served as a court-appointed expert witness in a patent dispute between the American Sperry Gyroscope Company and its German competitor, the Anschütz Company in Kiel. The founder of the German company was Hermann Anschütz-Kämpfe (1872–1931), a talented explorer and inventor who aspired to reach the North Pole by submarine. Since the direction of the Earth's magnetic field is almost vertical at the highest latitudes, the magnetic compass is useless and so Anschütz turned to the gyrocompass as a navigational tool. Einstein's testimony helped Anschütz prevail in court, and the two men subsequently became close friends.

The principle of the gyrocompass—a gyroscope that retains its orientation indefinitely—was first recognized by Léon Foucault in 1852, but it did not become a practical device until Anschütz's firm solved the problem of powering the gyroscope to keep it spinning in a housing that isolated it from the motion of the ship (or plane). Einstein contributed to the ultimate design of a practical gyrocompass and earned a royalty of 3 percent of the sales price. By the 1930s, Anschütz gyrocompasses had become the standard navigation device in most navies. It is not clear how Einstein reconciled his pacifist principles with his work on gyrocompasses during WWI, when they found wide use in German submarines.

Shortly after Rathenau's murder in 1922, the political tensions in Berlin ran so high that Einstein was tempted to leave Berlin and abandon academia. He wrote his friend Anschütz that Rathenau's death had traumatized him profoundly, and he proposed that he go to work as a technician in Anschütz's firm. The astonished Anschütz wrote to his friend the physicist Arnold Sommerfeld on July 12, 1922: "Here is the latest: Einstein is tired of Berlin and everything connected with it, the visits and official things, and *horribile scriptu*, he wants to go into technology." Although Einstein dropped his plan soon afterward, he retained his technical connection and his friendship with Anschütz. Following this strange episode, Anschütz offered Einstein a refuge in Kiel that was at his disposal at any time. It was an apartment on the ground floor of a house situated on the banks of the Schwentine River, near where it enters Kiel Bay. The apartment came completely equipped with a sailboat tied up at a jetty at the bottom of the garden. Einstein often made use of this refuge and brought his sons there for sailing holidays. Dieter Lohmeier and Bernhardt Schell, *Einstein, Anschütz und der Kieler Kreiselkompass* (Heide in Holstein: Verlag Boyens, 1992); also Albrecht Fölsing, *Albert Einstein: A Biography*, trans. E. Osers (New York: Viking, 1997).

11. Menasseh Meyer (1847?–1930) was born in Baghdad but was raised in the Baghdadi émigré community in Calcutta. He arrived in Singapore in 1873 and founded a company that prospered in the opium trade and other ventures. At one

time, he owned more Singapore real estate than anyone else. He was deeply religious and dominated the religious life of Singapore's Jewish community for over fifty years. In 1906 King Edward VII knighted him for his generosity and his role in civic affairs of the colony. For a detailed account of the Baghdadi Jews in Singapore, see Joan Bieder, *The Jews of Singapore*, ed. Aileen Lau (Singapore: Suntree Media, 2007).

12. Alfred Montor was a diamond merchant of German origin. He was the brother of the well-known stage and screen actor Max Montor, who appeared in the classic King Vidor film *Street Scene* (1931). Einstein commented that Alfred was also a gifted actor and that his wife was genuinely Viennese, even though she had grown up in Singapore. He clearly took a liking to both of them.

13. Einstein's passionate concern for the underprivileged is well known, but the following anecdote possibly less so. In Berlin, the Einsteins lived on the fourth floor of an apartment building that had the strict rule that only tenants could use the elevator, while domestic help had to use the back stairs. Einstein proposed to the doorman that he would use the back stairs, if the maid with her heavy grocery basket would then be allowed to ride the elevator. Carl Seelig, *Albert Einstein* (Zurich: Europa Verlag, 1960).

14. The atmosphere of the Jewish community center the Einsteins visited resembled that of the colony's British clubs, where Jews were not always welcome. It was a popular meeting place of Hong Kong's Jews. The original Jewish settlers, who had arrived soon after Hong Kong became a crown colony (in 1842), were Baghdadi Sephardim who retained close ties to Baghdad, Bombay, and Calcutta. Their entrepreneurial Diaspora brought many to Hong Kong, as well as to Shanghai, Singapore, and other Asian cities.

The Sassoon family, known as "the Rothschilds of the East," held sway over the religious life of the Hong Kong community. The synagogue services were held in Arabic. When European Ashkenazi Jews began to arrive in the 1880s, they attended the same synagogue, but the two groups mixed very little socially until the Jewish Community Center was founded early in the twentieth century. It provided a bridge between the two cultures. At the time of Einstein's visit, the center boasted a concert and lecture hall with a grand piano, as well as a library, a billiard room, and a bar presided over by a white-jacketed Chinese cocktail mixer. Caroline Plüss, "Sephardic Jews in Hong Kong: Constructing Communal Identities," *Occasional Papers of the Sino-Judaic Institute* 4 (2003): 57–79.

15. Following the announcement of Einstein's visit, at least eight books and five magazine articles dealing with relativity theory appeared in Japan, including four volumes of Einstein's collected papers and Atsushi Ishiwara's relativity treatise. Ezawa, "Impacts of Einstein's Visit."

16. Among the theoretical physicists who came to Kobe to welcome Einstein

were Atsushi Ishiwara (1881–1947), who had studied under Sommerfeld and Planck; Ayao Kuwaki (1878–1945), who had introduced the relativity theory to Japan in 1906 and had visited Einstein in Bern in 1909; and Hantaro Nagaoka (1865–1950), who belonged to the "first generation" of Japanese physicists and was best known for a Saturnian model of the atom, which he proposed in 1903. Nagaoka had been trained in Japan by the Scottish physicist C. G. Knott, but Ishiwara, Kuwaki, and another physicist, Kei-ichi Aichi (1880–1923), had all graduated from the University of Tokyo and had studied in Germany.

17. W. H. Solf, Report to the Foreign Office, Berlin. Tokyo, 3 January 1923. Siegfried Grundmann, *Einsteins Akte: Wissenschaft und Politik—Einsteins Berliner Zeit* (Berlin: Springer, 2004), pp. 231–35.

Wilhelm Heinrich Solf (1862–1936) was a philologist and diplomat who had worked in Germany's Colonial Office before WWI. He served as governor of Samoa from 1900 to 1910, resuming his diplomatic career under the Weimar Republic. After his death, the apartment of his widow, Johanna, was used as a clandestine meeting place of a group of anti-Nazi diplomats known as the Solf Circle. In 1943 the group was betrayed to the Gestapo, and several of its members were executed. Johanna Solf was arrested but survived the war.

18. Ishiwara was an established poet, as well as a highly regarded theoretical physicist. At about the time of Einstein's visit, he was forced to resign from Tohoku University because of the scandal surrounding his love affair with a poetess. Ishiwara's records of Einstein's lectures were later published, illustrated with Ippei Okamoto's cartoons. Ezawa, "Impacts of Einstein's Visit."

19. Grundmann, *Einsteins Akte*, pp. 231–35.

20. Sigfrid Berliner and Anna Berliner (1888–1977) returned to Germany in 1932. Being Jewish, in 1938 they emigrated again, this time to the United States, where they taught at various universities.

21. The contentiousness of Einstein's relationship with Heinrich Friedrich Weber (1843–1912) two decades earlier reflects the transition from classical to "modern" physics that occurred at the turn of the century. As a student, Einstein had been drawn to hands-on experimental work in Weber's well-equipped electrical laboratory, even though he considered Weber to be old-fashioned and accused him of "simply ignoring everything that came after Helmholtz." Einstein proposed an experiment to Weber to test the current "ether" hypothesis (similar, in principle, to the Michelson-Morley experiment), but Weber was unreceptive to the idea and told Einstein, "You are a very smart boy, but with one great fault, you won't let anyone tell you anything." For his part, Einstein was impertinent and persistently called him "Herr Weber" instead of "Herr Professor Weber." Fölsing, *Albert Einstein,* p. 79. Also Seelig, *Albert Einstein*, p. 48.

Ironically, this was not the only occasion on which the careers of the two men had intersected. Physicists had long ago observed that all "simple" solids (consisting of a single element) had approximately the same specific heat capacity (about 6 kcal/gram-mole.degree). This so-called Law of Dulong and Petit (1812) was explained much later by Boltzmann by relating the thermal energy (kT) to the vibrational energy of the constituent atoms. By extending Planck's concept of quantized radiation to the mechanical (vibrational) energy of atoms, Einstein predicted that at very low temperatures—that is, when kT was comparable to the vibrational energy— the specific heat would be lower than the classical Dulong-Petit value. The temperature dependence of the specific heat of diamonds predicted by Einstein's quantum theory turned out to be in excellent agreement with the experimental specific heat at low temperature—which had been measured thirty years earlier by none other than Friedrich Weber, then an assistant of Helmholtz. Albert Einstein, "Die Plancksche Theorie der Strahlung und die Theorie der spezifischen Wärme" [Planck's theory of radiation and the theory of specific heat], *Annalen der Physik* 22 (1907): 180–90.

22. Ezawa, "Impacts of Einstein's Visit," p. 3.

23. Unlike most Japanese universities, Waseda University was a private institution. It was founded in 1913 along Western lines by Shigenobu Okuma. Its mission was to "uphold the independence of learning, to promote the practical utilization of knowledge, and to create good citizenship."

24. Shifts of spectral lines caused by high temperature and electromagnetic fields were of interest to Einstein because they might mask the gravitational red shifts of solar spectra. Freundlich was even then attempting to measure the gravitational redshift of sunlight in the solar observatory in Potsdam.

25. A bold choice indeed. Beethoven's op. 47 is a brilliant work that puts considerable musical and technical demands on both the violinist and the pianist.

26. Ieyasu Tokugawa (1543–1616) was the founder of the Tokugawa shogun dynasty that ruled Japan from 1606 to 1868. This so-called Edo period, named after the shogunate's capital, Edo (now Tokyo), was the most peaceful period in the history of Japan. Fifteen thousand craftsmen worked for two years to construct his magnificent shrine, which is covered with gold leaf.

27. Albert Einstein, "My Impressions in Japan," trans. Hiroshi Ezawa, *AAPPS Bulletin* 15 (Oct. 2005): 20.

28. Leonor Michaelis (1875–1949) was professor at the University of Nagoya at the time of Einstein's visit. He was a physical chemist best known for elucidating the kinetics of enzyme reactions. After he immigrated to the United States, he was a professor at Johns Hopkins and, later, Rockefeller University.

29. Robert Koch (1843–1910), physician and pioneer of systematic bacteriology, was the discoverer of the anthrax and the tubercle bacterial organisms. He also studied the cholera microorganism and cholera epidemics, and he formulated rules for making water supplies safe.

30. The request to discuss relativity's genesis had come from the philosopher K. Nishida at Kyoto University. A record of the lecture appeared as Albert Einstein, "How I Discovered the Theory of Relativity (*Wie ich die Relativitätstheorie entdeckte*)," notes taken by Yun Ishiwara, trans. Y. O. Ono, *Physics Today* 15 (Aug. 1932): 45. Another translation of Ishiwara's notes, by Masahiro Morikawa, appeared in the *AAPPS Bulletin* 15 (Oct. 2005).

31. Todaiji was built in the year 752, and the temple's influence became so great that the capital was moved from Nara to Nagaoka for a time. It is still the largest wooden structure on Earth and is the site of many religious festivals.

32. Grundmann, *Einsteins Akte*, pp. 231–35.

33. The towns of Moji and Shimonoseki face each other across the narrow Kammon Strait between Japan's main island of Honshu and the southernmost island of Kyushu. Fukuoka, Miyake's hometown, is on Kyushu. The Mitsui Moji Club has since been converted into a luxury hotel, in which the Einsteins' room is preserved as a museum.

34. Tragically, both Hayasi Miyake and his wife were killed during WWII. After the war, their children asked Einstein to compose the epitaph for their gravestone.

3. HOMEWARD BOUND: PALESTINE AND SPAIN (1923)

1. The *Haruna Maru* carried cargo as well as passengers. Built in 1922, she was brand new, and with a displacement of 10,500 tons, she was only a little larger than the *Kitanu Maru*. In 1942, the *Haruna Maru* was wrecked in a fog and sank.

2. Einstein hoped to extend Eddington's approach, itself based on the work of Hermann Weyl (1885–1955) to a unified field theory. (This is the same Sir Arthur Eddington who measured the deflection of starlight by the sun.) His mathematical approach to finding a connection between gravitation and electromagnetism looked so promising to Einstein that a few days later, he wrote an article presenting his ideas on Eddington's theory. He mailed the manuscript to Planck for submission to the Prussian Academy when the *Haruna Maru* arrived in Port Said. Albert Einstein, "Zur allgemeinen Relativitätstheorie" [On the general relativity theory], *Sitzungsberichte der Königl. Preuss. Akad. Wissenschaften* (1923): 32–38 and 76–77.

When Eddington's theory failed to lead to verifiable physical results, Einstein lost interest in it.

3. It is interesting that when Einstein delivered his Nobel lecture in Stockholm in July 1923, he chose to speak not about the photoelectric effect but about the theory of relativity. Albrecht Fölsing, *Albert Einstein: A Biography*, trans. E. Osers (New York: Viking, 1997), pp. 535–51.

4. Albert Einstein Dupl. Archive, Princeton University, Call Number 08-074.

The articles by Bohr that accompanied Einstein to Japan were probably the trilogy "On the Quantum Theory of Line Spectra," in which Bohr described the generalized model of the hydrogen atom and proposed the shell structure of atoms, with electrons occupying "stationary states" characterized by quantum numbers. He had arrived at this insight less by mathematical reasoning than by an inspired amalgamation of the spectroscopic and chemical properties of the elements of the periodic table. Despite the model's successes, Bohr was well aware of its empirical nature and of its internal inconsistencies, and he knew that classical physics could not resolve them.

These deficiencies of the "old quantum theory" would soon be swept aside by the discovery of quantum mechanics. Abraham Pais, *Niels Bohr's Times* (Oxford: Clarendon Press, 1991), p. 192.

5. Alluding to Portia, the heroine in Shakespeare's *Merchant of Venice*.

6. Possibly, one of the Veddas, the aboriginal inhabitants of Sri Lanka.

7. Beside the well-known circus, the Hagenbeck family also owned and operated the famous Hagenbeck *Tierpark* (Animal Park). Opened in 1907, it was the first park in which animals were confined without fences.

8. Einstein's chiding of the French for having learned little since the Napoleonic era, and his condemnation of the occupation, must be understood in the context of German and French politics in the aftermath of the war. Einstein recognized that the occupation would not only cripple the German economy, but also greatly benefit the resurgent ultranationalists. France, for her part, had suffered enormously during the war, and the French government was under intense pressure from the Right to keep her neighbor to the east impotent. Poincaré insisted on Germany keeping up her war reparations payments, as required by the Treaty of Versailles; whereas Britain and the United States were willing to see the debts reduced, to help get Germany back on her feet, and they took no part in the occupation. Predictably, the French army's uncompromising military occupation led to passive resistance by the populace and to acts of sabotage. Several saboteurs were executed and became Nazi martyrs.

9. Among those who came aboard Einstein's train in Lod were Benzion Mossinsohn, principal of the Jaffa Gymnasium, and Menachem Ussishkin, head of the Jewish National Fund. The secular philosopher Asher Ginsberg (1856–1927), who opposed "political" Zionism, had a happy reunion with Einstein at the railway station in Jerusalem.

10. The building was constructed between 1907 and 1910 and served as a hospice for German pilgrims until the British administration acquired it during WWI. It was named after Wilhelm's wife and is known as the "Victoria Augusta" to this day. It now serves as a hospital for mainly Arab patients.

11. Sir Herbert Louis Samuel (1870–1963), later Viscount Samuel, was

Britain's first Jewish cabinet minister and helped shape the Balfour Declaration. Through his appointment, he became the first Jew to govern that territory in two thousand years.

12. The meeting between Einstein and Samuel blossomed into a lasting friendship, and while their correspondence over thirty years ranged far and wide, their particular shared interest was the philosophy of science. Aside from being a politician— he was a member of parliament, home secretary, and leader of the Liberal Party— Samuel also authored several books. Einstein praised Samuel's book *Belief and Action: An Everyday Philosophy* (Indianapolis: Bobbs-Merrill, 1937), but chided Samuel gently for selling short "the atomic physicists and possibly also the biologists, who adhere to the materialistic principle," deeming it possible to understand, say, life processes in terms of physical laws. At a time when scientists had almost no comprehension of the fundamental life processes, Einstein's letter included this prescient passage:

"With regard to the attitude physics takes towards life processes, it is true that restricting ourselves to the concepts and laws of physics does not allow us to attain rational comprehension of the over-all process. Maybe we human beings will never succeed in this. But one must not conclude from this *that physics does not, in principle, encompass the life-processes.* That would be a bankruptcy declaration by physics, which is not compelled by any necessity." United Kingdom Parliamentary Archive, Letter Ref. No. SAM/E/32.

In his memoir, Samuel painted this portrait of Einstein: "Recognized everywhere as the greatest scientist of our age, he carries his immense fame without the smallest self-consciousness, without either pride or diffidence. When this illustrious man, urbane and gentle, was expelled from his university professorship by the Nazis, uprooted from his home and robbed of all his property, he never condescended to rancor, or resentment, or repining. His demeanor had the dignity and detachment of a civilized man, a scientist and a philosopher, confronted by barbarians." Herbert Louis Samuel, *Grooves of Change: A Book of Memoirs* (Indianapolis: Bobbs-Merrill, 1946), p. 301.

13. Arthur Ruppin (1876–1943) was a German-born sociologist who moved to Palestine in 1907; he pioneered Jewish settlements. The philosopher Samuel Hugo Bergmann (1883–1975), whom Einstein first met when he was a professor in Prague, started the Library of the Hebrew University. He later became the university's first president.

14. From a collection of Helen Bentwich's letters published by her daughter, Jenifer Glynn. It appears here with her kind permission. Jenifer Glynn, *Tidings from Zion* (London: I. B. Tauris, 2000), p. 95.

Helen Bentwich, neé Franklin, came from a well-established Anglo-Jewish family who traveled widely and often. She was an experienced and dedicated social worker and organizer, and she took an active interest in Palestine and in the development of Jewish settlements. Her letters offer a remarkable insider's view of the prevailing economic and cultural climate in Palestine, of Arab-Jewish relations, and of Samuel's governance.

15. The Bentwiches were a very musical family, and in the Mozart quintet, Norman's two sisters, Margery and Thelma, played violin and cello, respectively. Mr. Feingold must have provided the second viola.

16. Tel Aviv was initially founded in 1910, but the Turks evacuated the Jews living there in 1916. After the British administration took over in 1917, the town grew rapidly, drawing most of its early Jewish inhabitants from the nearby ancient city of Jaffa, particularly after the 1921 riots in that city. At the time of Einstein's visit, Tel Aviv had about twenty thousand inhabitants.

17. Hermann Struck (1876–1944) (Hebrew name Hayyin Aharon ben David) was born and educated in Berlin. He was highly regarded for his evocative etchings and lithographs. Struck wrote the standard work on graphic technique and was a teacher of Chagall, Liebermann, and Corinth. As an early Zionist, he moved to Haifa in 1923 and worked there for the remainder of his life.

Einstein sat for a striking lithograph portrait by Struck on this visit (see the first page of photo insert).

18. The Technikum was not officially opened until 1924, and then only after intense debates over whether lectures were to be given in German or in Hebrew. Aharon Czerniawski (1887–1966) taught physics at the Reali School and the Technion, and Elias Auerbach (1882–?) was a physician and a noted historian. Both were early German Zionists who had settled in Haifa.

19. Nahalal, the first workers' settlement (moshav ovedim), was founded in 1921 by veteran pioneers, some of them former members of the first kibbutz, Deganyah. Eighty families of settlers received twenty-five acres of land each. Their first task was to drain the malaria-infested swamps, which had thwarted two previous attempts to start a settlement there. The layout of Nahalal as devised by the architect Richard Kaufmann became the pattern for many subsequent moshavim. It was based on concentric circles, with the communal buildings (school, administrative, and cultural offices, shops and warehouses) located in the center, the homesteads in the innermost circle, farm buildings in the next, and gardens, orchards, and fields in the outermost circle.

20. A few years later, Einstein expressed his mixed feelings about Palestine succinctly: "I love that land, but am afraid of its chauvinism." Quoted in Carl Seelig, *Albert Einstein* (Zurich: Bertelsmann-Europa, 1960), p. 318.

21. The monumental statue of Ferdinand de Lesseps, his arm stretched out in welcome, stood at the Canal's very entrance. It was demolished by the Egyptians following the French–British occupation of Port Said in 1956, and today only its massive, empty pedestal remains.

22. The *Oranje* was an old and small ship, built in 1903. Her displacement was 4,400 tons, and she had a top speed of fifteen knots. She was returning from the Dutch East Indies when Einstein and Elsa came aboard in Port Said. Later that same year, she was sold to a French company and was renamed *Anfa*. She was scrapped in 1936.

23. Einstein's sparse diary accounts of his activities in Spain have here been supplemented by contemporary reports: Thomas F. Glick, *Einstein in Spain: Relativity and the Recovery of Science* (Princeton: Princeton University Press, 1988), pp. 100–149; Siegfried Grundmann, *Einsteins Akte: Wissenschaft und Politik—Einsteins Berliner Zeit* (Berlin: Springer Verlag, 2004), pp. 243–49.

Glick's book covers many details of Einstein's visit, as well as the scientific and political background to his visit; also, the content of Einstein's three Madrid lectures.

24. Esteban Terradas (1883–1950) played a key role in bringing modern physics to Spain and had been the first to raise the idea of inviting Einstein to Spain. Einstein was at first reluctant—because he could not speak Spanish—and postponed the trip several times. Terradas was held in particularly high esteem by Einstein. Glick, *Einstein in Spain*, p. 118.

25. Glick, *Einstein in Spain*, p. 104.

26. Ilse von Tirpitz was the eldest daughter of Alfred von Tirpitz (1849–1930), who had been charged by Wilhelm II with strengthening the Imperial battle fleet in order to rival the British navy. In WWI, von Tirpitz favored unrestricted submarine warfare against neutral shipping. When restrictions were imposed on that policy, he resigned.

Ilse was married to the diplomat Ulrich von Hassell (1881–1944), who originally supported Hitler but later became disillusioned with him and joined the resistance movement. Following the failed plot to assassinate Hitler on July 20, 1944, von Hassell was arrested by the Gestapo. After he was tried and convicted as a conspirator, he was executed.

27. Blas Cabrera (1878–1945) was an experimental physicist who, along with Terradas, played an important role in introducing relativity theory to Spain.

28. Einstein's first cousin Lina was married to Julio Kocherthaler. She was not only Einstein's cousin, but also a close friend, and the two corresponded frequently with each other. Julio's brother, Kuno, was an industrialist with holdings in Spain and was married to the art historian María Luisa Cazurla.

29. These reports appeared in the monarchist newspaper *ABC* on March 6 and 10, 1923. Glick, *Einstein in Spain*, p. 130.

30. The Toledo party included, apart from the two Kocherthaler brothers with their wives, the noted El Greco scholar Manuel B. Cossío (1857–1935) and the Republican essayist and philosopher José Ortega y Gasset (1883–1955).

31. The painting depicts the soul of Count Orgaz ascending to heaven, assisted by an angel, while his physical body is being lowered into his coffin by Saint

Stephen. Painted in 1586 by El Greco, it is located in the vestibule of the Church of Santo Tome. The two medieval synagogues Einstein visited were the Tránsito and the Santa María la Blanca, erected as a synagogue in 1180 and now a museum.

32. Ernst Langwerth von Simmern (1865–1941) was a well-traveled and experienced diplomat. It is remarkable that Einstein was on such friendly terms not only with Langwerth but with all the German consular officials he encountered on his travels. They all sent enthusiastic reports to the Foreign Office in Berlin, extolling Einstein's character and modesty, and affirmed that his visit had greatly advanced the German cause. Grundmann, *Einsteins Akte,* pp. 243–49.

33. Einstein had reason to be impressed. The violinist was Antonio Fernández Bordas (1870–1950), a student of Pablo de Sarasate and a celebrated virtuoso.

34. While the physicist Jerónimo Vecino had lectured on relativity and had instigated Einstein's visit to Saragossa, Einstein's scientific interests were more closely aligned with those of Antonio de Gregorio Rocásolano (1873–1941), who worked on the Brownian motion of colloids. Einstein visited him in his laboratory.

35. Emil von Sauer (1862–1942) was a student of Nikolai Rubinstein and Franz Liszt. He was a composer, as well as a highly regarded piano virtuoso.

4. SOUTH AMERICA (1925)

1. Philipp Frank, *Einstein: His Life and Times* (New York: Alfred A. Knopf, 1947), p. 205. Frank reports that Einstein was disturbed by the sight of faces devoid of understanding while he talked of things that were close to his heart.

2. In classical Boltzmann statistics, particles are counted as if distinguishable from each other. The young Indian mathematician Saryendra Nat Bose recognized that this is unjustifiable in the case of photons. He derived Planck's radiation formula by treating photons as a gas of *indistinguishable* particles, and he sent his manuscript to Einstein, who translated it into German and had it published. Einstein recognized that the same considerations applied to atoms as well as to photons, and he generalized Bose's statistical approach to create a quantum theory of ideal gases, now known as Bose-Einstein statistics. Albert Einstein, "Quantentheorie des idealen Gases," *Königl. Preuss. Akad. der Wissenschaften* (1925): 18–25.

The theory allowed Einstein to make two remarkable predictions: (1) the viscosity of liquefied gases would disappear at sufficiently low temperatures; and (2) at very low temperatures, the wave-functions of atoms overlap and atoms condense into an aggregate. The first phenomenon, superfluidity of helium, was observed by Willem Hendrik Keesom in 1928, and the second, the existence of so-called Bose-Einstein condensates, was demonstrated in 1995 by experimenters at MIT and at the University of Colorado.

3. The initial contract with Einstein was made by the Asociación Hebraica, a Jewish cultural group, but in response to Einstein's wish, the official invitation was issued by the University of Buenos Aires. A considerable portion of the funds, $4,000 plus free passage for Einstein and his wife, were provided by the Asociación Hebraica and other Jewish organizations. Additional funds were raised for Einstein's visits to Uruguay and Brazil. Eduardo L. Ortiz, "A Convergence of Interests: Einstein's Visit to Argentina in 1925," *Ibero-Amerikanisches Arkiv, Zeitschrift für Sozialwissenschaften und Geschichte, Berlin* 21 (1995): 67–126.

4. Among those seeing Einstein off were Elsa; Moritz Katzenstein (1872–1932), a celebrated surgeon and close friend; and Alexander Bärwald (1877–1930), a noted Berlin architect who emigrated to Palestine later in 1925 and taught at the Technion in Haifa.

5. Richard Robinow (1875–1945), Marie's husband, came from a family of distinguished Jewish citizens devoted to public service in Hamburg. After the Nazis came to power, he was initially allowed to continue his law practice because of his service in WWI, but in 1938 he was arrested by the Gestapo. Freed from a concentration camp, he managed to emigrate to London with his family.

6. The twenty-three-thousand-ton *Cap Polonio* was 662 feet in length and had a top speed of eighteen and a half knots. She was a classic black-hulled ship with a straight bow and three funnels, and she boasted a tiled swimming pool. Built in 1914, she served as an auxiliary cruiser during WWI and was surrendered to Britain as war reparation in 1919. The Hamburg South America Line purchased her back in 1921, and she was scrapped in 1935.

7. Carl Jesinghaus (1886–1948) was a professor of philosophy and psychology in Argentina since 1912. He went back to Germany and a professorship in Würzburg from 1935 till 1945, returning to Argentina after the war.

8. Else Jerusalem, neé Kotányi (1877–1942), was a well-known progressive feminist and a successful author. One of her novels saw thirty-two editions. Born in Vienna of Jewish parents, around 1910 she moved to Buenos Aires with her second husband, Victor Widakowich, who was professor of embryology at La Plata University. She converted to Protestantism in 1911. We know from Einstein's diary that he addressed her as 'Panther Cat' in person, as well as in his diary entries.

9. Émile Meyerson (1859–1933) was a chemist, epistemologist, and philosopher of science. He was an opponent of positivism and conducted an extensive correspondence with Einstein. Einstein probably read *La déduction relativiste* (Paris: Payot, 1925), and while he credited the author for being clever, he objected to his notion that the work of Weyl and Eddington had any bearing on the theory of relativity.

The Russian-born social philosopher David Koigen (1879–1933) taught in Berlin. Koigen's manipulation of abstract concepts seemed curious to Einstein, who speculated that it might be more meaningful for a person from the East. David Koigen, *Der moralische Gott* [The ethical God] (Berlin: Jüdischer Verlag, 1922).

10. Einstein's diary entries while in Argentina were supplemented with details from other sources, particularly Ortiz, "A Convergence of Interests." Ortiz discusses several aspects of Einstein's visit that are not mentioned here—e.g., the heated philosophical debate that ensued following Einstein's visit.

11. Mauricio Nirenstein (1877–1935) taught Spanish literature and was administrative secretary of the University of Buenos Aires. It was useful that he also had ties to the Sociedas Hebraica Argentina, the Jewish organization that provided considerable financial support for Einstein's visit.

After Einstein left Argentina, Nirenstein wrote an article about a conversation with Einstein in which he discussed the epistemology of science, which was a controversial subject at the time. Ortiz, "A Convergence of Interests," p. 105. For the philosophical background to Einstein's visit, and a manuscript of Einstein's introductory lecture at the university, see Alejandro Gangui and Eduardo L. Ortiz, "Einstein's Unpublished Opening Lecture for His Course on Relativity Theory in Argentina, 1925," *Science in Context* 21, no. 3 (2008): 435–50.

12. Leopoldo Lugones (1874–1938), a prolific and influential writer and social philosopher, was initially an anarchist, then a socialist, and finally a fascist. He had great interest in science and wrote a pamphlet about relativity. He and Einstein had met in Geneva on a League of Nations commission. Lugones first suggested inviting Einstein to Argentina when there was concern about Einstein's safety following Rathenau's murder. Lugones's writings ranged over novels, fantastic stories, Argentine history, and linguistics. In 1938, despairing and disillusioned, Lugones committed suicide by taking cyanide.

13. Einstein actually wrote he had had a 'nose full' (*Neese pleng*) of mass meetings from New York during the 1921 fund-raising trip, using a colloquialism common in Berlin.

14. Alfredo Hirsch (d. 1956) was partner and later president of the gigantic grain-exporting firm Bunge y Born. He was a prosperous and generous philanthropist and patron of the arts.

15. Richard Gans (1880–1954) had been among those who argued against inviting Einstein to La Plata because of Einstein's pacifist politics. He suggested that the physiologist Emil Abderhalden be invited instead. Ortiz, "A Convergence of Interests," p. 84. Oddly, Abderhalden is best known as the discoverer of the fictitious "protective enzymes" and as one of the most egregious scientific mountebanks. See Walter Gratzer, *Eurekas and Euphorias* (Oxford: Oxford University Press, 2002), pp. 317–19.

Gans was an avid German nationalist, and when he was offered a professorship in Königsberg in 1925, he returned to Germany. As a Jew, he was dismissed from his post when the Nazis came to power in 1933. Remarkably, he survived the Nazi era and WWII by working on a weapons project for which his expertise in magnetism was deemed essential. Gans's survival against all odds was facilitated by a seemingly silent conspiracy of German physicists who attested to the feasibility of the futile project: the use of X-rays generated by batteries of betatrons (electron accelerators) to bring down Allied bombers. Gans worked on the project throughout the war, then returned to Argentina in 1947 and taught in La Plata and Buenos Aires. Pedro Waloschek, *Todesstrahlen als Lebensretter* [Death rays as life savers] (Norderstedt: Books on Demand, 2004); also Edgar Swinne, "Richard Gans: Hochschullehrer in Deutschland und Argentinien," *Beiträge zur Geschichte der Naturwissenschaften und der Technik* 14 (1992).

16. As well as a prominent physician, José Arce (1881–1968) was a politician and diplomat. He served in the Chamber of Deputies for many years; under Juan Perón's government he went into exile in Madrid and the United States.

17. Coriolano Alberini (1886–1960), professor of philosophy at the universities of Buenos Aires and La Plata, was an opponent of "American pragmatism," which he perceived as a threat to traditional religious values. He was the only one of his travel companions who was highly regarded by Einstein, and the two men corresponded with one another for several years. Butty, an engineering professor, was interested in the philosophical implications of relativity and followed Bergson's ideas. Argentina's foreign minister, Angel Gallardo, was another travel companion on this journey.

Missing from Einstein's account of his stay in Córdoba is any mention of the well-known peace activist and physician Georg Nicolai (1874–1964), who was then professor of physiology at Córdoba University. He and Einstein had known each other well in Berlin, and in 1914 the two men had issued the "Manifesto to Europeans," in response to the infamous "Appeal to the Cultured World." They had found few signers. Subsequently, Nicolai and Einstein had had a serious personal falling-out that involved Einstein's older stepdaughter, Ilse Einstein. Walter Isaacson, *Einstein: His Life and Universe* (New York: Simon and Schuster, 2007), pp. 243–46.

18. In Einstein's original German: '*Lakierte Indianer, skeptisch-zynisch ohne Kulturliebe, im Ochsenfett verkommen.*'

19. Karl Gneist (1868–1939). Siegfried Grundmann, *Einsteins Akte: Wissenschaft und Politik—Einsteins Berliner Zeit* (Berlin: Springer, 2004), pp. 254–56.

20. Einstein explained that Else Jerusalem was 'broges' with him, a Yiddish expression meaning "peeved."

21. The poem for Else Jerusalem, in translation, read: "This is for the Panther Cat / although she furiously went to hide / in the jungle harsh and wild / she gets this picture, nonetheless."

22. Einstein referred to these luminaries—rectors, deans, ambassadors—as *Grosskopferte,* literally, "big-headed ones," a humorously derogatory colloquialism.

23. Carlos Vaz Ferreira (1872–1958) was a philosopher and the author of numerous books. He is noted for introducing liberal and pragmatic ideas into South American society.

24. Grundmann, *Einsteins Akte*, p. 258.

25. Fridtjof Nansen (1861–1930) was an outstanding athlete, scientist, diplomat, and explorer whose feats of endurance in the far North are legendary. As head of the Norwegian mission to the League of Nations at the end of WWI, he effected the relief and repatriation of hundreds of thousands of refugees and prisoners of war. As High Commissioner of Refugees, he invented the Nansen passport for stateless persons. He also helped save millions of lives during the Russian famine (1921), in the Greek–Turkish war (1922), and by the resettlement of the remnants of Armenians (1925). He was awarded the Nobel Peace Prize in 1922.

26. The *Valdivia,* built in Marseilles in 1911, had a gross weight of seven thousand tons and a top speed of fifteen and a half knots. From 1915 to 1919 her owners lent her to the British Admiralty for use as a hospital ship. She was scrapped in 1933.

27. Einstein's diary entries while in Brazil are here complemented by other sources, particularly A. T. Tolmasquim and I. C. Moreira, "Einstein in Brazil: the Communication to the Brazilian Academy of Sciences on the Constitution of Light," *History of Modern Physics, Proc. XXth Intl. Congress of History of Science* (Brepols: Turnhout, 1997), pp. 229–42; also Thomas F. Glick, "Between Science and Zionism: Einstein in Brazil," *Episteme* 9 (1999): 101–120. Einstein's visit was sponsored jointly by the Brazilian Academy of Science, the Polytechnic School, and the University of Rio de Janeiro.

28. The earliest proponents of these theories were George-Louis Leclerk, Comte de Buffon (1707–1788) and Cornelius De Pauw (1739–1799).

29. Aloysio de Castro (1881–1959) was chairman of the Faculty of Medicine (*Richtiger Affe*) in Rio. The author was the feminist poet Rosalina Coelho Lisboa (1900–1975).

30. Einstein's comments in the original German are included in Tolmasquim and Moreira, "Einstein in Brazil." An English translation appears in Richard A. Campos, *Still Shrouded in Mystery: The Photon in 1925:* http://arxiv.org/ftp/physics/papers/0401/0401044.pdf

Compton had observed that in collisions between a photon and an electron, the photon's frequency is lowered, while the total energy and momentum are conserved. Arthur H. Compton, "A Quantum Theory of the Scattering of X-rays by Light Elements," *Phys. Rev.* 21 (1923): 483–502. The statistical interpretation of the Compton Effect was proposed by N. Bohr, H. A. Kramers, and J. C. Slater in "The Quantum Theory of Radiation," *Phil. Mag.* 47 (1924): 785–802. The statistical explanation offered by this so-called BKS theory was disproven experimentally in a "coincidence experiment" that showed that energy and momentum were conserved in each individual photon-electron collision, not just statistically. Walther Bothe and Hans Geiger, "Über das Wesen des Comptoneffekts: ein experimenteller Beitrag zur Theorie der Strahlung" [Regarding the nature of the Compton Effect: An experimental contribution to radiation theory], *Z. Phys.* 32 (1925): 639–63.

The adherents and opponents of the BKS theory were led by two revered physicists, Bohr and Einstein, and the debate between the two factions was a passionate one. For an account of this fascinating controversy, see Abraham Pais, *'Subtle Is the Lord . . .' The Science and the Life of Albert Einstein* (Oxford: Oxford University Press, 1982), pp. 416–22.

31. Hubert Knipping (1868–1937). Grundmann, *Einsteins Akte*, p. 259.

32. Cândido Mariano da Silva Rondon (1865–1958) founded the Indian Protection Service in 1910. It is among the most humane of such organizations.

33. Shortly after Einstein's departure from Rio, the adherents to Auguste Comte's positivism, led by Professor Licinio Cardoso, launched an attack on Einstein and the relativity theory. Cardoso's paper "Imaginary Relativity" precipitated a fierce debate at the academy, which ended with the rout of the Cardoso faction. Glick, "Between Science and Zionism: Einstein in Brazil," 101–120.

34. The Hamburg South American Line's *Cap Norte* had a displacement of fourteen thousand tons and a top speed of fourteen and a half knots. Her distinction was that she was not coal-fired but oil-fired, and (supposedly) smoke-free. In the early weeks of WWII, she was intercepted and captured by HMS *Belfast*; thereafter she served as an Allied troopship.

35. Einstein and Elsa's fellow dinner guests on this occasion included the French diplomat Roland de Margerie, Countess Hedwig Berta Sierstorpff, the liberal politician and editor of the *Berliner Tageblatt* Theodor Wolff, the lawyer Hugo Simons (immortalized in Otto Dix's portrait), the industrialist Emile Mayrisch, and the French author Jean Schlumberger. Harry Graf Kessler, *Tagebücher 1918–1937* (Berlin: Deutsche Buch-Gemeinschaft, 1967), p. 456.

36. Kessler, *Tagebücher*, p. 396.

5. NEW YORK AND PASADENA (1930–1931)

1. Dieter Lohmeier and Bernhardt Littow, *Einstein, Anschütz und der Kieler Kreiselkompaß* [Einstein, Anschütz and the Kiel gyrocompass] (Heide in Holstein: Verlag Boyens, 1992). This book contains the text of many letters dealing with technical and personal matters that were exchanged between Einstein and Anschütz until Anschütz's death in 1931. The letters bear out Einstein's detailed involvement with the daunting electrical and mechanical engineering problems presented by a prac-

tical gyrocompass, and they testify to the high regard in which the two men held each other. See also Albrecht Fölsing, *Albert Einstein: A Biography*, trans. E. Osers (New York: Viking, 1997), p. 520, 542, 596.

2. Among the flaws of the new theory were that Planck's constant, *h*, did not appear in it and that it was inconsistent with relativity theory. In January 1932, Einstein wrote to Pauli: "You were right, after all, you rascal, you." Fölsing, *Albert Einstein*, pp. 602–607.

3. Thomas Levenson, *Einstein in Berlin* (New York: Bantam, 2003), p. 381.

4. Fölsing, *Albert Einstein*, p. 451.

5. Einstein first met Belgium's Queen Elizabeth (1876–1965) in the previous year while visiting his uncle Cesar Koch in Antwerp, when he played trios with her and a lady-in-waiting. Elizabeth belonged to the royal Bavarian Wittelsbach family. She was the niece of the ill-fated Elizabeth of Austria, Emperor Franz Josef's wife. She and Einstein corresponded with each other until the end of his life. Fölsing, *Albert Einstein*, p. 630.

6. Arthur Henry Fleming (1856–1944) was a major benefactor of Caltech during the school's early days.

Robert Andrews Millikan (1868–1953) is best known for his "oil drop experiment" in which he determined the charge of the electron. At the time of Einstein's visit, Millikan was president of Caltech while continuing his research on the nature of cosmic rays. Millikan had met Einstein in 1921 and had initially invited him in 1925, but Einstein's commitment to visit South America and, later, his 1928 heart attack delayed his acceptance.

7. Edwin Hubble (1889–1953) announced his dramatic discovery of galaxies far beyond the Milky Way in 1925. To be sure, the existence of many galaxies beyond the Milky Way had been contended by the brilliant William Herschel more than a century before Hubble, based on his own visual observations with telescopes of his own design. Richard Holmes, *The Age of Wonder: The Romantic Generation and the Discovery of the Beauty and Terror of Science* (New York: Vintage Books, 2010), p.123.

Among the other scientists with whom Einstein interacted in Pasadena were Paul Epstein (1883–1966) and Theodore von Kármán (1881–1963).

8. The SS *Belgenland*, built as a cargo ship in Belfast in 1914, was refitted as a troopship when WWI broke out. She had a displacement of twenty-seven thousand tons, was powered by three screws, and sported three funnels. In 1923, she was outfitted as a luxury liner with a tiled swimming pool, but during the Depression she was no longer profitable, and she was broken up in 1936.

The Nazis cited Einstein's preference for the British-owned *Belgenland* over a German liner as evidence of his disloyalty to Germany. In fact, he chose the *Belgenland* because she called at both New York and Los Angeles.

9. Theodor Koch-Grünberg (1872–1924) was a distinguished German ethnolo-

gist who spent many years studying the Amazon Indians and collecting their legends.

10. When the *Belgenland* arrived in San Diego, journalists interviewed several of her passengers, who reported that Einstein would occasionally stroll absent-mindedly into the dining salon clad in his pajamas. *The New York Times*, January 1, 1931.

11. In Berlin, Toni Mendel, the widowed mother of Hertha Mendel, had been a close friend of Einstein's for a long time. The two often sailed together, attended concerts together, and together they read and discussed Freud's latest writings. Both left Germany in 1932, but they continued to correspond with each other into the 1950s while she lived in Canada. Einstein refers to the licorice candy as *Bärendreck*, a colloquialism that literally means "bear turds."

12. Einstein observed that R, the ratio of the lowest to the highest weight registered by the scale, was 2/3. Let g be the acceleration due to gravity, and let a be the ship's vertical acceleration. Making use of the equivalence of gravitational and inertial forces, Einstein noted that $R = (g - a) / (g + a)$, so that $a = g(1 - R) / (1 + R)$. Since $R = 2/3$ and $g \sim 10$ m/sec^2, Einstein was able to estimate the acceleration to be $a = (1/5)g \sim 2$ m/sec^2.

13. *The New York Times*, December 10–16, 1930. Some of the dating in the travel diary's transcript is erroneous. The dates given here are based on newspaper accounts of Einstein's activities in New York.

14. *The New York Times*, December 12, 1930.

15. Russian-born Abraham Menachem Mendel Ussishkin (1863–1941) was an engineer who became an important Zionist leader. He promoted the use of the Hebrew language, the creation of agricultural settlements, and the Hebrew University, and he served as president of the Jewish National Fund from 1923 until his death. He and Einstein had met previously in the course of the Einsteins' 1923 visit to Palestine.

16. Felix M. Warburg (1837–1937) was a member of the Hamburg banking family. He was a brother of the bankers Max and Paul Warburg, and of the renowned art historian Aby (Abraham Moritz) Warburg. Felix was an investment banker who came to the United States in 1894. He was a well-known philanthropist and a supporter of Zionist causes.

17. Megaera (literally, "the jealous one") is one of the three Furies of Greek mythology. It is not clear why Einstein conferred this epithet on Kreisler's wife, Harriet, but he was apparently not alone in his point of view.

18. Rabindranath Tagore (1861–1941) was a renowned Bengali writer, poet, musician, and graphic artist who had many enthusiastic followers in Europe in the aftermath of WWI. Many perceived him and Einstein to be representatives of the Eastern and Western cultures. Their two conversations have been preserved and published. The first conversation took place in Einstein's Berlin flat in 1926, and the

second, in the home of Toni, Bruno, and Hertha Mendel in Wannsee, in July 1930. Rabindranath Tagore, *Das goldene Boot*, ed. M. Kämpfchen, trans. A. O. Carius (Düsseldorf: Artemis & Winkler, 2005), pp. 547–55.

The transcript of their conversations suggests that the two men agreed on little. See Wendy Singer, "'Endless dawns' of imagination," *The Kenyon Review* 23, no. 2 (Spring 2001): 7–33; also Dipankar Home and Andrew Robinson, "Einstein and Tagore: Man, Nature and Mysticism," *Journal of Consciousness Studies* 2, no. 2 (1995): 167–79.

19. Paul A. M. Dirac (1902–1984) made many important contributions to theoretical physics between 1925 and 1930: the most important ones were his derivation of the relativistic wave equation and his theory of the electron, which predicted the existence of positrons—positively charged particles with the same mass as electrons.

20. Frau Kuhn almost certainly told Einstein and Elsa of Henry Morgan's arriving in Panama as an indentured laborer and becoming captain of a pirate fleet following his release. After capturing the Spanish fort of San Lorenzo on Panama's Atlantic coast, he crossed the isthmus with five hundred of his buccaneers and sacked the original Panama City on the Pacific coast. Until then, the Spaniards had shipped Peruvian gold to Panama City, then overland to the Atlantic port of Porto Bello, and from there, to Spain. Once the hostilities between Britain and Spain ended, Morgan returned to England a wealthy man and was knighted.

21. The theoretical physicist Richard Tolman (1881–1948) acted as Einstein's interpreter and guide at Caltech on numerous occasions. Tolman's model envisaged the universe expanding until gravity causes its collapse (big crunch) and another big bang. The simplicity and symmetry of his model are appealing, but modern observations have shown that the most distant galaxies are *not* slowing down; on the contrary, they are speeding up.

22. Albert Abraham Michelson (1852–1931), an ingenious experimenter who made important contributions to interferometry, is best known for the Michelson-Morley experiment (1887) with Edward Wilson Morley. The experiment demonstrated convincingly that the velocity of light was independent of Earth's motion in space. Surprised at the result, they repeated the experiment with ever greater precision, but with the same result. It sounded the death knell of ether theories and prepared the scientific community for Einstein's relativity theory—even though Einstein himself was unaware of the Michelson–Morley experiment at the time he proposed the relativity principle.

23. Walter Sidney Adams (1877–1956), the director of the Mount Wilson Observatory, identified the companion star of Sirius as a white dwarf.

The gravitational redshift of Sirius B was of interest because it was expected to be large and provided another test of general relativity theory. (The gravitational redshift is separate from the Doppler redshift of light emitted by a receding source.)

Eddington, who regarded general relativity as intuitively evident, used the reported gravitational redshift to verify his estimates of the star's mass and radius. Modern work has shown that his estimates were erroneous, as was the determination of the gravitational redshift. For the long and tangled history of redshift measurements, see Jay B. Holberg, *Sirius, Brightest Diamond in the Night Sky* (Berlin: Springer Verlag, 2007), p. 141.

24. The astronomer Charles Edward St. John (1857–1935) is best known for his work on sunspots and the solar atmosphere.

25. Carl Lämmle (1867–1939) was born in Laupheim, a short distance from Ulm, Einstein's birthplace. He immigrated to the United States when he was seventeen, and after practicing several professions, he became an independent film producer in Hollywood and, later, president of Universal Studios. Among the films he produced are *The Hunchback of Notre Dame*, *Phantom of the Opera*, and *Frankenstein.*

Lämmle remained strongly attached to his Swabian hometown, returning to it often and sending food and money to alleviate the town's hardships following WWI. After the Nazis came to power, he could no longer visit Germany, but he provided affidavits that made it possible for some three hundred Jews to escape to America. After WWII, Lämmle was made an honorary citizen of Laupheim, where streets, a square, and the high school are named after him.

26. Abraham Hoffman, "Albert Einstein at Caltech," *Calif. Hist.* 76, no. 4 (1997/8): 108–21.

27. The footage Einstein viewed showed the pilgrimage of Indians to the shrine of Our Lady of Guadalupe, Mexico's most important saint, and the bullfighting scenes, which disgusted Einstein thoroughly. In the end, Stalin obliged Eisenstein to return to the Soviet Union, but the film footage remained with Sinclair. Eisenstein was never able to see it. Eventually, some of it was edited and released under the title *¡Que viva México!*

28. The participants in the dinner and the radio broadcast were Albert A. Michelson, Robert A. Millikan, William W. Campbell, Walter S. Adams, Charles E. St. John, Edwin P. Hubble, and Richard C. Tolman.

29. This observation was erroneous. The huge magnetic field associated with a sunspot is now thought to emerge as an enormous magnetic field loop that reenters the sun's surface at a different sunspot. This explains why the magnetic fields in pairs of sunspots, as observed by St. John, rotate in opposite directions.

30. *The New York Times*, February 8, 1931.

31. The country house of Samuel Untermeyer (1858–1940) in Palm Springs is now a luxury hotel known as The Willows.

32. The twenty-one-thousand-ton *Deutschland* was a luxuriously appointed liner on the Hamburg–New York run. Requisitioned by the Reichsmarine (the German navy) in 1940, she was sunk in 1945 near Lübeck in a British air attack.

33. *The New York Times,* March 5, 1931; also Fölsing, *Albert Einstein*, p. 640.

34. *The New York Times*, March 15, 1931.

6. BERLIN AND OXFORD (1931)

1. Albert Einstein, "Die Plancksche Theorie der Strahlung und die Theorie der spezifischen Wärme" [Planck's theory of radiation and the theory of specific heat], *Annalen der Physik* 22 (1907): 180–90.

Frederick Alexander Lindemann (1886–1957) was aware of the disastrous situation that German Jews faced, and he played a leading role in securing academic posts in Britain for many German Jewish scientists in the 1930s. Among the most prominent of these were Franz Simon (later Sir Francis Simon), Leo Szilard, Kurt Mendelsohn, Fritz London, Heinrich Kuhn, and Nicholas Kurti. During the Second World War, Lindemann was a close associate and scientific advisor of Winston Churchill, and for his services he was ennobled to Baron and, subsequently, Viscount Cherwell.

2. James Clerk Maxwell's (1831–1879) great achievement was the electromagnetic field theory, which encompasses all electric and magnetic phenomena and is summarized in Maxwell's Equations. Expressing a law of nature in terms of a continuous "field" represented, in Einstein's view, the most fruitful advance in physics since Newton. It opened the way to relativity theory, as well as to quantum mechanics. Albert Einstein, *Mein Weltbild* (Zurich: Europa Verlag, 1934), p. 186.

3. Erich Ludendorff realized too late what a mistake it had been to support Hitler. In 1933, after Hindenburg turned the government over to Hitler, Ludendorff accused Hindenburg of having "delivered our holy German Fatherland to one of the greatest demagogues of all time" and added, "I solemnly prophesy that this accursed man will cast our Reich into the abyss and bring our nation to inconceivable misery. Future generations will damn you in your grave for what you have done." Ian Kershaw, *Hitler 1889–1936: Hubris* (New York: Norton, 1998), p. 427.

4. The Einstein Foundation was funded by a group of wealthy German industrialists. The architect of the modernistic Einstein Tower was Erich Mendelsohn (1887–1953), who eventually immigrated to the United States. For a detailed account of the Einstein Foundation and of Freundlich's troubled relationship with both Einstein and Hans Ludendorff, see Klaus Hentschel, *The Einstein Tower*, trans. A. M. Hentschel (Stanford: Stanford University Press, 1997).

5. Max von Laue (1879–1960) made numerous important contributions to physics. He is best known for demonstrating in 1912 that the recently discovered X-rays were an electromagnetic radiation of short enough wavelength to be diffracted by crystals. He launched X-ray diffraction as an experimental technique, which, in

time, developed into one of the most powerful tools of physics and chemistry, as well as molecular biology.

6. After Hitler's takeover, Lindemann arranged for Erwin Schrödinger (1887–1961) to obtain a fellowship at Oxford's Magdalene College, but after a short time there, Schrödinger returned to Austria in 1936—to the considerable annoyance of Lindemann and Einstein. Schrödinger was a professor at the University of Graz, and after Hitler annexed Austria in 1938, he felt impelled to publish an abject, repentant apology for past political errors in which he called on Austrian scientists to support the Nazi regime. He was, nevertheless, dismissed from his professorship, though he managed to escape from Austria to Italy in the summer of 1938. He received an appointment at the Institute for Advanced Studies in Dublin and spent the war years there. Walter Moore, *Schrödinger: Life and Thought* (Cambridge: Cambridge University Press, 1989), pp. 320–51.

7. Albert Einstein, "Zum kosmologischen Problem der allgemeinen Relativitäts-theorie" [On the cosmological problem of general relativity theory], *Proceedings of the Prussian Academy of Sciences* (1931): 235–37. For a more comprehensive, yet nontechnical discussion of the history of the cosmological constant and of other "mistakes" of Einstein's, see Steven Weinberg, *Lake Views: This World and the Universe* (Cambridge, MA: Harvard University Press, 2009), pp. 186–95.

8. Käthe Dorsch (1890–1957) was a celebrated stage and film star. Gotthold Ephraim Lessing (1729–1781) was an associate of Voltaire and was an important early author of the Enlightenment. *Minna von Barnhelm* is a comedy about an officer whose sense of honor comes in conflict with his love for a woman.

9. Ernst Barlach (1870–1938) was originally a patriotic supporter of WWI and volunteered for service at the front. He returned from the war a confirmed and activist pacifist. Under the Nazis, his antiwar sculptures were confiscated as degenerate art, and he was forbidden to work.

10. Roger Highfield and Paul Carter, *The Private Lives of Albert Einstein* (London: Faber & Faber, 1993), p. 207.

11. In his response to Freud's birthday greetings, Einstein wrote that while he (Freud) had slipped under the skin of so many people, he had never had an opportunity to slip under his. Freud told Einstein that he was fortunate because only physicists could criticize his (Einstein's) theories, while *his* theories were criticized by one and all. (Princeton AE Archives, Freud to Einstein, 4/29/1931. Einstein to Freud 3/22/1929)

12. His host for this house concert was Berthold Israel, of the huge Berlin department store Nathan Israel. The Israel family had been invited to settle in Berlin by Frederick the Great and was one of the oldest Jewish families in the city. The family was known for combining economic success with social responsibility.

13. Alexander Moszkowski (1851–1934) was the older brother of the prominent composer and pianist Moritz Moszkowski. When his book was about to be pub-

lished, Einstein's friends, particularly Max and Hedi Born, were concerned that Einstein's enemies would use it to brand him as a publicity seeker, and they urged Einstein to forbid its publication. The book eventually appeared under the title *Einstein the Seeker* (not the original *Conversations with Einstein*), and, while it is flawed, it does contain some interesting material, such as Einstein's denial that the energy in the formula $E = mc^2$ can be liberated in practice. But others, e.g., Hermann Weyl, considered this a possibility as early as 1918. (Highfield and Carter, *Private Lives*, 198–200. For an exhaustive account of the Moszkowski affair, see http://www.mathpages.com/home/kmath630/kmath630.htm)

14. A year later, Einstein's younger son, Eduard, visited his aunt Maja for a week and spent much of the time sitting at this piano, playing Mozart "with the greatest accuracy and soberness." But soon after this visit, Eduard's condition worsened. He entered the Burghölzli psychiatric clinic, where Einstein visited him for the last time in 1933. Eduard was to remain at the Burghölzli, off and on, until his death in 1965.

15. The German mathematician Emil Julius Gumbel (1891–1966) published in 1931 a detailed account of the numerous political murders committed by Nazis. Gumbel was a member of the German League for Human Rights, and in 1932 he lost his position at the University of Heidelberg because of his anti-Nazi activities.

16. Albert Ballin's (1857–1918) story is one of rags to riches and, ultimately, tragedy. He rose from modest beginnings to become the managing director of the Hamburg America Line (Hapag), owners of the largest merchant fleet of Wilhelmine Germany. Wilhelm II, as well as Ballin, recognized that large passenger liners were symbols of German imperial glory. Ballin invented commercial cruise ships and cruising, and under him, at the turn of the century, Hapag provided affordable passage (steerage) to the thousands of emigrants passing through Hamburg on their way to America. He also provided rail travel and housing for them. Ballin was a staunch German nationalist and was the only nonconverted Jew on a friendly footing with the kaiser. When Germany collapsed at the end of WWI, Wilhelm II abdicated and fled to the Netherlands; on the same day, Albert Ballin committed suicide. Bernhard Huldermann, *Albert Ballin*, trans. M. J. Eggers (London: Cassell, 1922).

The SS *Albert Ballin* was built in 1922 and had a displacement of twenty-one thousand tons. She was in regular service on the Hamburg–Southampton–New York run until WWII. After 1933, the ship's name was anathema to the Nazis and Hapag was obliged to rename her the SS *Hansa*. She was sunk by a mine in the Baltic in 1945. After she was salvaged by the Soviet Union, she was renamed *Sovietsky Sojus* and was based in Vladivostok until 1981.

17. John G. Griffith, "Albert Einstein at Winchester 1931," *The Trusty Servant* 62 (1986): 5. This incident was kindly drawn to the author's attention by Freeman Dyson.

18. Edward Arthur Milne (1866–1950) is best known for his work on the thermodynamics of stars, at a time when the energy source (nuclear fusion) of stars was not yet understood. Milne considered the kinematic relativity theory, published in 1935, his most important work, but it is now largely forgotten.

19. Adolf Georg Wilhelm Busch (1891–1952), celebrated German violinist and composer, opposed Hitler and moved to Switzerland in 1927. Although he was not Jewish, he renounced Germany in 1933, and after emigrating to the United States, he and his equally famous pianist-partner, Rudolf Serkin (1901–1993), were founders of the Marlboro Music School and Festival near Brattleboro, Vermont.

20. Some additional details about Einstein's quartet partners: When Marie Soldat (1863–1955) was sixteen, she played a recital in the Austrian village of Pörtschach, where Brahms was spending the summer. He was in the audience and was so taken by her talent that he persuaded Joseph Joachim to audition her. She became a distinguished soloist and founded an all-women string quartet. Brahms was so delighted by her performance of his Violin Concerto at its Vienna premiere (in 1885) that the next evening he took her to the *Prater*, Vienna's amusement park, and to the theater.

Marie Soldat first met the Deneke family in 1899 when the family still lived in London, and she stayed in their home whenever she gave a concert in London. In her memoir, Margaret recalled that Soldat's concert dresses were bought for her in Vienna by Clara Wittgenstein, Ludwig Wittgenstein's aunt. For additional details, see Michael Musgrave, "Marie Soldat 1863–1955: An English Perspective," in *Beiträge zur Geschichte des Konzerts,* ed. S. Kross, R. Emans, and M. Wendt (Bonn: Gudrun Schröder Verlag, 1990), pp. 319–30; also Max Kalbeck, *Johannes Brahms* (Berlin: Deutsche Brahms Gesellschaft, 1908–21), 4th ed., vol. 3, p. 158; also Styra Avins, *Johannes Brahms: Life and Letters* (Oxford: Oxford University Press, 1997), p. 591.

Erna Schulz (1887–1938) was another brilliant student of Joachim, winning high praise as a violin soloist before joining the Wietrowetz Quartet as violist. She resigned from the quartet in 1912 and settled in London, where she taught and played in concerts.

The Welshman Arthur Williams was the cellist of the celebrated Klingler Quartet, successor to the Joachim Quartet. He was a student of Robert Hausmann, Brahms's favorite violoncellist and member of the Joachim Quartet. At the outbreak of WWI, Williams was interned and returned to Britain.

21. This famous 1742 violin had been bought for Soldat by Louis Wittgenstein, Ludwig's uncle.

22. Margaret Deneke, *What I remember* (Unpublished typescript), Deneke Deposit, Box 14, Bodleian Library, University of Oxford. Margaret Deneke (1882–1969) was a dedicated benefactor of Lady Margaret Hall (LMH), Oxford's first

women's college, raising substantial funds in lecture-concerts that she presented in England and the United States. In 1931 LMH elected her to an honorary fellowship.

23. Paul Kent, "Einstein at Oxford," *Oxford Magazine* (Michaelmas Term, 2005): 8–10.

24. John Sealy Edward Townsend (1868–1957) had been the professor of experimental philosophy since 1900. He is today remembered for his work on electrical discharges and for his early attempts to measure the charge of the electron. It is interesting that Einstein noted in his diary that the German physicist Emil Wiechert (1861–1928) was the first person to measure the charge-to-mass ratio of the electron. Did Townsend tell him that? Wiechert performed his experiment at almost the same time as J. J. Thomson, who was awarded the Nobel Prize for it.

25. The venue for the widely anticipated debate between Einstein and Philipp Lenard at the scientific congress was moved from Frankfurt to the rural Bad Nauheim in order to avoid political and anti-Semitic demonstrations. The confrontation became notorious: the debate itself was low-key, but the bitterness between the two men, both well-regarded physicists and Nobel Prize winners, remained. For a perceptive account of the personality and the politics of Lenard and his ally, Johannes Stark, see Walter Gratzer, *The Undergrowth of Science* (Oxford: Oxford University Press, 2000), pp. 244–66.

In 1938, after the Nazis had come to power, Lenard published his own account of the Nauheim discussion. He confessed to having treated "the Jew" as if he was a proper Aryan human being, as was customary at that time. This had been wrong, he asserted, but he excused his behavior by noting that the relevant racial guidelines were not published until 1922. For a detailed account of the Bad Nauheim episode, see Charlotte Schönbeck, *Albert Einstein und Philipp Lenard: Antipoden im Spannungsfeld von Physik* (Berlin: Springer Verlag, 2000); also Walter Isaacson, *Einstein: His Life and Universe* (New York: Simon & Schuster, 2007), pp. 284–89.

26. Wilhelm Blaschke (1885–1862) was a leading geometer and occupied a professorial chair in Hamburg at the time.

27. For an account of Freundlich's uneasy relations with Einstein, see Hentschel, *The Einstein Tower*, pp. 98–102.

7. RETURN TO PASADENA (1931–1932)

1. Only Tomáš Masaryk, the president of Czechoslovakia, bothered to reply to Einstein. He did so thoughtfully, in his own hand. Thomas Levenson, *Einstein in Berlin* (New York: Bantam Books, 2003), p. 410.

2. The MS *Portland* provided regular freight and passenger service between Europe and the US Pacific coast. Built in 1928, she was 480 feet in length and had

a top speed of fifteen knots. She featured an outdoor swimming pool, and her passenger accommodations were luxurious for the time. She was scuttled by her crew off the Azores in 1943 in order to prevent her capture by the Allies.

3. Einstein may have known that it had been shown long ago that Euclid's constructions can be performed with only a compass. These so-called Mascheroni constructions were discovered independently by Georg Mohr (1640–1697) and Lorenzo Mascheroni (1750–1800).

Euclid's axioms and geometrical constructions (e.g., the construction of an equilateral triangle), ca. 300 BCE, arguably mark the birth of mathematics. They were taught to all students of mathematics for the next 2,300 years.

4. In Leiden, Einstein was particularly close to the physicist Paul Ehrenfest (1880–1933), who was a student of Boltzmann, and to his physicist wife Tatiana Ehrenfest-Afanaseva (1876–1964). Both made important contributions to statistical mechanics.

5. The theory of Oswald Veblen (1880–1960) and Banesh Hoffmann (1906–1986) employed a five-dimensional geometry, which was related to Einstein's latest approach to finding a unified field theory.

6. Adriaan Fokker (1887–1972) was a student of Lorentz and is best known for deriving the Fokker–Planck equation, which has important applications in cosmology.

7. Adolf Gustav Smekal (1895–1959) predicted the Raman Effect on theoretical grounds. It is the basis of Raman spectroscopy, which is used, for example, to investigate rotational and vibrational states of molecules.

8. In Antony van Leeuwenhoek's (1632–1723) simple (single-lens) microscopes, the specimen is mounted on a screw and is critically positioned in front of a tiny, hand-ground lens. Using such primitive tools, with which it is estimated that he attained magnifications up to 500x, Van Leeuwenhoek discovered and described many microscopic life-forms. His discoveries were widely circulated and had a profound effect on the field of biology.

9. Heike Kamerlingh-Onnes (1853–1926) was professor of experimental physics in Leiden from 1882 to 1923. He was the first person to liquefy helium and is the discoverer of superconductivity.

10. Egon Friedell, *Kulturgeschichte der Neuzeit: Die Krisis der europäischen Seele von der schwarzen Pest bis zum Weltkrieg* [Cultural history of the current era: The crisis of the European soul, from the Black Death to the World War], 3 Vols. (Munich: Beck, 1927–31).

Egon Friedell (1878–1938) was a cultural historian, as well as actor, cabaretist, essayist, theater critic, and journalist. His *Cultural History* was outlawed by the Nazis. Following the Nazi occupation of Austria, Friedell jumped to his death before he could be arrested..

11. Max Born (1882–1970) was an old friend of Einstein's whose book had recently been published. Max Born and Pascual Jordan, *Elementare Quantenmechanik* [Elementary quantum mechanics] (Berlin: Springer, 1930).

12. The Swiss psychoanalyst Carl Gustav Jung (1875–1961) claimed to have been strongly influenced by Chinese wisdom—specifically, through Richard Wilhelm's German translation of the *I Ching* and the Taoist text *The Secret of the Golden Flower*. Einstein read Jung's 1929 commentary on the latter.

13. Heinrich Hoffmann's *Der Struwwelpeter* (translated as *Slovenly Peter* by Mark Twain) was published in 1844. Since then, almost all German-speaking children have been familiar with the book's illustrations and its often sadistic stories.

14. Following the 1930 election for the Reichstag, the Nazis had sufficient strength to boycott the Weimar parliament, obliging Chancellor Heinrich Brüning (1885–1970) to govern by emergency decree. During the short life of the Weimar Republic, Brüning was its longest-serving chancellor (1930–1932).

15. Josef Kastein, *Eine Geschichte der Juden* (Berlin: Ernst Rowohlt, 1931). Kastein (1890–1946) was an ardent German Zionist whose writings expressed his faith in the mission and destiny of the Jewish people. He immigrated to Palestine in 1935.

16. *Los Angeles Times*, January 1, 1932.

17. The deuterium nucleus contains a neutron as well as a proton and thus has approximately twice the mass of an ordinary hydrogen nucleus (a single proton). The isotope occurs naturally, with an abundance of about 0.02 percent. Harold C. Urey (1893–1981) and his collaborators reported their discovery in 1931, and the issue of the *Physical Review* containing their article had just arrived at Caltech at the start of Einstein's 1932 visit. Urey was awarded the Nobel Prize in 1935.

18. Franz Simon (1893–1956), later Sir Francis Simon, soon left Berlin for good, as did Einstein and other German Jewish scientists. He was invited to Oxford University, where he worked in low-temperature physics. He was knighted for his war work during WWII.

Moritz Schlick (1882–1936) studied physics but became an influential philosopher and exponent of logical positivism. He considered any statement that cannot be proven or disproven to be metaphysical and, hence, meaningless. He was the founder of the Vienna Circle, and, while he was not Jewish, he was critical of Hitler. In 1936 a former student shot him dead on a staircase of Vienna University. The killer was convicted and sentenced, but after the Nazi occupation of Austria, he was released and joined the Nazi Party.

19. Albert Einstein and Willem de Sitter, "On the Relation between the Expansion and the Mean Density of the Universe," *Proc. Nat. Acad. Sci. (USA)* 18 (1932): 223–24. The paper tentatively implies the existence of an initial singularity—i.e., the big bang.

20. Tom Mooney (1882–1942) was a union organizer and member of the Inter-

national Workers of the World (IWW). By the use of blatantly false evidence, Mooney was convicted of planting a bomb during a demonstration in San Francisco in 1916. His became a cause célèbre, and many intellectuals and public figures in Europe and the United States signed appeals on his behalf. Their efforts finally succeeded in 1939, when the newly elected Democrat governor of California, Culbert Olson, ordered his release. When he emerged from prison, Mooney was greeted by twenty-five thousand of his supporters.

21. Jacob Gould Schurman (1854–1942) was a philosopher before he turned diplomat. As a visitor at Caltech, like Einstein, he stayed at the Athenaeum.

22. From the track of an ionizing particle in a magnetic field, the particle's charge and mass can be determined. The cloud chamber was invented in 1912 by the Scottish physicist C. T. R. Wilson (1869–1959). After undergoing many improvements, it was used by Arthur Compton (1892–1962) to demonstrate the so-called Compton Effect: when X-rays are scattered by an electron, their wavelengths are *lengthened* by an amount that depends on the scattering angle. The Compton experiment was thought to provide the best evidence that radiation was quantized. Wilson and Compton shared the Nobel Prize for physics in 1927.

23. Jesse W. M. DuMond (1892–1976) was a brilliant instrument builder, best known for his curved crystal spectrometers for diffracting X-rays and gamma rays—analogously to light diffracted by a grating.

24. Lili Petschnikoff, *The World at Our Feet* (New York: Vantage Press, 1968). Born in Chicago, Petschnikoff studied in Berlin under Joseph Joachim and performed to great acclaim as violinist and violist in Europe and America—often with her husband, the celebrated violinist Alexandre Petschnikoff (1873–1949). After divorcing him, she settled in Hollywood, where Einstein often played chamber music with her during his three sojourns in Pasadena. She was a close friend of Elsa and corresponded with her frequently.

In a remarkably insightful letter Elsa wrote from Princeton in 1934, she chided Lili severely for having, apparently, expressed some sympathy for the new Nazi regime. Elsa told her of the brutal conditions under which Jews lived, that Albert had fought against the Nazis with every fiber of his heart, and that Jews were not the only victims in the developing tragedy, which was unmatched in history. She also reminded Lili that her former husband and her friends, Bruno Walter and Lotte Lehmann, were Jews—as were her own children.

Letter, *Elsa Einstein to Lili Petschnikoff*, dated February 5, 1934. Image posted on ECHO (European Cultural Heritage Online), http://nausikaa2.mpiwg-berlin.mpg.de/cgi-bin/toc/toc.test.cgi?dir=100_G;step=thumb.

25. Oswald Veblen (1880–1960) made important contributions to geometry and the theory of functions. He and Einstein would soon become colleagues at Princeton.

26. Einstein had met the Spanish liberal philosopher José Ortega y Gasset in the

course of their one-day excursion to Toledo in 1923 (see chapter 3). The historian and philosopher Oswald Spengler (1880–1936) proposed a cyclical theory of history and promoted a nationalist version of socialism. He was originally an intellectual hero of the Nazis, but when he rejected their racial policy in 1933 he was ostracized.

27. Long before he was the director of the Manhattan Project, J. Robert Oppenheimer (1904–1967) was a highly regarded theoretical physicist at the University of California in Berkeley. Because of his interest in cosmic rays and neutron stars, he had strong ties to Caltech and the Mount Wilson Observatory. Shortly after his Caltech lecture, he submitted a paper proposing the neutron's existence. Abraham Pais, *J. Robert Oppenheimer: A Life* (Oxford: Oxford University Press, 2006), p. 26.

James Chadwick (1891–1974), a student of Ernest Rutherford, was awarded the Nobel Prize in physics in 1934 for his discovery.

28. The service was held in the Scott Methodist Episcopal Church and was attended by 750 people. Julius Rosenwald (1862–1932) was a former president and part-owner of Sears, Roebuck, and Company and was the chief architect of the company's success. He was an associate of Booker T. Washington and a director of the Tuskegee Institute. His foundation donated many millions to philanthropic causes, including the construction of five thousand rural "Rosenwald schools" in the southern United States.

29. The song was "Behüt dich Gott, es wär zu schön gewesen" ("God Be with You, It Would Have Been Too Beautiful, but It Was Not to Be"). It was first heard in the 1884 opera *Der Trompeter von Säckingen* (*The Trumpeter from Säckingen*) and remained popular into the 1930s.

30. The MS *San Francisco* was built in 1927 and, like the *Portland*, was powered by diesel engines. She was scuttled by her crew in 1943 to prevent her capture by the Allies.

8. OXFORD, PASADENA, AND LAST DAYS IN EUROPE (1932–1933)

1. Philipp Frank, *Einstein: His Life and Times* (New York: Alfred Knopf, 1947), p. 226. The *Junkers* were members of the Prussian nobility.

2. Margaret Deneke, Unpublished, hand-written notebook, 1932. Deneke Deposit, Box 25, Bodleian Library, University of Oxford.

3. Deneke, Unpublished notebook, 1932.

4. Abraham Flexner (1866–1959) was an influential educator and played an important role in restructuring medical education. The funds for the Institute for Advanced Study ($5 million) were donated by the department-store owner and philanthropist Louis Bamberger and his sister, Caroline Bamberger Fuld.

5. Frank, *Einstein*, pp. 268–71.

6. *The New York Times*, December 6, 1932; December 7, 1932.

7. Frank, *Einstein*, p. 226.

8. The *Oakland,* built in 1929, was requisitioned by the German navy at the start of WWII and was renamed *Sperrbrecher IV* (blockade runner). She sank following an air attack on Brest in 1944—a fate similar to that of several of the ships Einstein had traveled on.

9. *The New York Times*, December 11, 1932.

10. Paul Langevin (1872–1946) made many important contributions to quantum physics and was an early champion of relativity theory in France. He was also a prominent supporter of anti-fascist and antiwar causes.

11. Franz Oppenheimer, *Weder so, noch so: Der dritte Weg* (Potsdam: Alfred Protte Verlag, 1933). Oppenheimer (1864–1943) was a prolific and ardent writer on economical and social issues. The book Einstein read was his last publication before the Nazi takeover. In it, he argued that a surplus of labor was at the root of the social upheaval.

12. Paul Dirac (1902–1984) had introduced his relativistic wave equation four years earlier. The Dirac equation, in which the state of a particle is expressed in terms of 4x4 matrices (spinors), provides a description of particles with spin 1/2 (e.g., electrons) that satisfies both relativity theory and quantum mechanics. The equation allowed Dirac to predict the existence of antiparticles: particles of identical mass and spin, but with opposite electrical charge. His prediction was confirmed in 1932 by Carl Anderson's discovery of the positron (the antiparticle of the electron)—one of the triumphs of theoretical physics.

13. Kurt von Schleicher (1882–1934) was a former staff officer in the *Reichswehr* and an adept politician. As chancellor, he tried to form a coalition between rightist unions and Gregor Strasser's dissident Nazi faction. The attempt failed, and in June 1934, in the course of the so-called Röhm-Putsch, he and his wife were murdered (as was Strasser) by Hitler's SS. Hitler thereby eliminated his immediate predecessor, a legally appointed chancellor.

14. Einstein's stay in Pasadena was funded by the Oberlaender Trust of the Carl Schurz Memorial Foundation. *Los Angeles Times*, January 10, 1933.

15. Einstein read B. L. van der Waerden, *Moderne Algebra* (Berlin: Springer, 1930). Bartel Leendert van der Waerden (1903–1996) was an influential mathematician and historian of mathematics.

16. *Los Angeles Times*, January 23, 1933.

17. After he was awarded a Guggenheim Fellowship in 1926, Linus Carl Pauling (1901–1994) studied quantum mechanics in Europe under Sommerfeld, Bohr, and Schrödinger. On his return, he used quantum mechanics to investigate the electronic structure of atoms and molecules, and he pioneered the field of quantum chemistry. In later years, he also made contributions to molecular biology and was a

political activist. He was awarded the Nobel Prize for chemistry in 1954 and the Nobel Peace Prize in 1962.

18. Theodore von Kármán (1881–1963) was born in Hungary, studied in Göttingen, and made many significant contributions to aerodynamical science.

19. Irving Fisher (1867–1947) was the first American celebrity economist and the first mathematical economist. He was also a eugenicist and a proponent of the justly ridiculed theory of "focal sepsis," which contended that mental illness is caused by infectious material in recesses of the bowel and the roots of teeth. Einstein probably read his *Booms and Depressions: Some First Principles* (New York: Adelphi, 1932).

20. Ralph Fowler (1889–1944) held the chair in theoretical physics at the Cavendish Laboratory. He wrote a seminal work (with Arthur Milne) on the spectra and the thermodynamics of stars. He was a mentor of Paul Dirac and was married to Ernest Rutherford's daughter, Eileen.

21. *Los Angeles Times*, March 3, 1933.

22. *Los Angeles Times*, March 12, 1933

23. The Communist Reichstag delegates had already been arrested or gone underground by the time the empowerment law was passed. The law was to have run only until April 1, 1937, but it was extended repeatedly and remained the "legal basis" of Hitler's government until 1945.

For a comprehensive account of how Hitler achieved total control of the government, see Ian Kershaw, *Hitler 1889–1936: Hubris* (New York: Norton & Co, 1998), pp. 469–526.

24. For detailed accounts of Einstein's messy separation from the Prussian Academy of Sciences and other academic institutions, see Albrecht Fölsing, *Albert Einstein* (New York: Viking, 1997), pp. 661–65.

25. *The New York Times*, September 9 &10, 1933. Fölsing, *Albert Einstein*, pp. 674–76.

26. Walter Isaacson, *Einstein: His Life and Universe* (New York: Simon & Schuster, 2008), p. 419.

27. *The New York Times,* October 18, 1933; Isaacson, *Einstein*, p. 425–47; Fölsing, *Albert Einstein*, pp. 679–92.

EPILOGUE (1933–1955)

1. More details of Einstein's life in America are found in Albrecht Fölsing, *Albert Einstein*, trans. E. Osers (New York: Viking, 1997), pp. 679–92; Walter Isaacson, *Einstein: His Life and Universe* (New York: Simon and Schuster, 2007), pp. 425–47; Philipp Frank, *Einstein: His Life and Times* (New York: Alfred Knopf, 1947), pp. 265–98.

2. Isaacson, *Einstein*, pp. 428–31.

3. *The Royal Gazette and Colonist Daily*, May 28, 1935.

4. Isaacson, *Einstein*, p. 437.

5. Known as "the EPR paper" after its authors' last initials, it refers to A. Einstein, B. Podolsky, and N. Rosen, "Can Quantum-Mechanical Description Be Considered Complete?" *Phys. Rev.* 47 (1935): 777–80. For the debate surrounding the EPR paper, see Abraham Pais, *'Subtle is the Lord . . .' The Science and the Life of Albert Einstein* (Oxford: Oxford University Press, 1982), p. 454; also Albrecht Fölsing, *Albert Einstein* (New York: Viking, 1997), pp. 696–99.

Very briefly: the article suggested that "quantum entanglement" between two particles at a distance was unrealistic; since then, however, entanglement has been demonstrated experimentally.

6. When massive bodies undergo vehement motion, gravitational radiation is emitted. Several highly sophisticated experiments to detect gravitational waves are currently underway. Albert Einstein and Nathan Rosen, "On Gravitational Waves," *J. Franklin Soc.* 223 (1937): 43–54.

A. Einstein, L. Infeld, and B. Hoffman, "Gravitational Equations and the Problem of Motion" Part 1: *Annals Math.* 39 (1938): 65–100; Part 2: A. Einstein and L. Infeld. *Annals Math.* 41 (1940): 455–64.

7. Fölsing, *Albert Einstein*, pp. 693–705.

8. For detailed accounts of Einstein's role in the atomic bomb project, and his political activities after the war, see Fölsing, *Albert Einstein*, pp. 706–41.

9. Planck had suffered greatly. He lost two daughters in childbirth and one son in WWI before his other son, in the waning days of WWII, was executed by the Nazis after being implicated in the July 20 plot to assassinate Hitler.

10. Letter, Albert Einstein to Toni Mendel, dated: Princeton, 24 March 1948. Letter in private possession.

11. Letters, to and from Toni Mendel, 15 June 1953. Albert Einstein Archive, Box/Folder 41-254, 41-255.

12. Frank, *Einstein*, p. 297.

Select Bibliography

Calaprice, Alice. *The Einstein Almanac*. Baltimore: Johns Hopkins Press, 2005.

Fölsing, Albrecht. *Albert Einstein: A Biography*. Trans. E. Osers. New York: Viking, 1997.

Frank, Philipp. *Einstein: His Life and Times*. New York: Knopf, 1947.

Glick, Thomas F. *Einstein in Spain: Relativity and the Recovery of Science*. Princeton: Princeton University Press, 1988.

Herneck, Friedrich. *Einstein Privat*. Berlin: Buchverlag Der Morgen, 1978.

Grundmann, Siegfried. *Einsteins Akte: Wissenschaft und Politik-Einsteins Berliner Zeit*. Berlin: Springer Verlag, 2004.

Highfield, Roger, and Paul Carter. *The Private Lives of Albert Einstein*. London: Faber and Faber, 1993.

Isaacson, Walter. *Einstein: His Life and Universe*. New York: Simon and Schuster, 2007.

Kershaw, Ian. *Hitler: 1889–1936: Hubris*. New York: W. W. Norton, 1998.

Levenson, Thomas. *Einstein in Berlin*. New York: Bantam Books, 2003.

Neffe, Jürgen. *Einstein: A Biography*. Trans. S. Frisch. New York: Farrar, Straus & Giroux, 2005.

Pais, Abraham. *'Subtle is the Lord . . .' The Science and the Life of Albert Einstein*. Oxford: Oxford University Press, 1982.

———. *Niels Bohr's Times, In Physics, Philosophy, and Polity*. Oxford: Clarendon Press, 1991.

———. *Einstein Lived Here*. Oxford: Clarendon Press, 1994.

Schilpp, Paul A., ed. *Albert Einstein, Philosopher-Scientist*. New York: Tudor Publ. Co., 1949.

Seelig, Carl. *Albert Einstein*. Zurich: Bertelsmann-Europa, 1960.

Stern, Fritz. *Einstein's German World*. Princeton: Princeton University Press, 1999.

Index